INTERNATIONAL CENTRE FOR MECHANICAL SCIENCES

COURSES AND LECTURES - No. 167

J.D. ACHENBACH
NORTHWESTERN UNIVERSITY
EVANSTON, ILLINOIS

A THEORY OF ELASTICITY WITH MICROSTRUCTURE FOR DIRECTIONALLY REINFORCED COMPOSITES

SPRINGER-VERLAG WIEN GMBH

Originally published by Springer-Verlag Wien New York in 1975

ISBN 978-3-211-81234-1 ISBN 978-3-7091-4313-1 (eBook)
DOI 10.1007/978-3-7091-4313-1

PREFACE

The material presented in this volume is part of a set of notes for a series of lectures, which was given under the title "Mechanical Behavior of Directionally Reinforced Composites under Dynamic Loading Conditions," at the International Institute of Mechanical Sciences, Udine, Italy, in July 1973. Since the printed volume is mainly devoted to mathematical modeling of the mechanical behavior of directionally reinforced composites, in particularly by means of theories of elasticity with microstructure, the present title of the volume is more appropriate.

For a laminated medium, a homogeneous continuum model whose mechanical behavior is described by a theory of elasticity with microstructure, was proposed in 1966, jointly by the author and G. Herrmann. Since then, the basic ideas have been extended to theories for fiber-reinforced composites, and to viscoelastic material behavior, as well as to theories to describe thermal effects, and effects of large deformations and nonlinear mechanical behavior. For laminated media and fiber-reinforced composites the development of the theory is presented in Part II. To place these new theories, and their areas of application, in a proper perspective, the well-established effective modulus theory is discussed in some detail in Part I.

The research work in the area of the mechanical behavior of composite materials reported here, was carried out under the sponsorship of the Office of Naval Research under Contract ONR 064-483 with Northwestern University. This support is gratefully acknowledged.

Evanston, October 1973

J.D. Achenbach

CONTENTS

PART I

THE EFFECTIVE MODULUS THEORY

CHAPTER 1

INTRODUCTION

The concept of compounding reinforcing elements and a matrix material to form a directionally reinforced composite material has become widely accepted in materials engineering. As an example we mention fiber-reinforced composites which are now used increasingly in technological applications. The formulation of an adequate theory to describe the mechanical behavior of directionally reinforced composites., especially for dynamic loading conditions, poses an interesting challenge to workers in the field of continuum mechanics.

1.1. Fiber-Reinforced Composites

Fiber-reinforced composites have at least two distinct parts: the high-modulus oriented filaments and the low-modulus matrix. The fibers can range from the more conventional glass to the somewhat more exotic boron and graphite. For the matrix a wide range of epoxies are used for applications at lower temperatures, while metals are employed at higher temperatures. Composites normally are manufactured in two forms – as tape or combined with metal in structural shapes. The tapes are employed to build up multiple layers or laminations of collimated high-strength filaments. Filaments within these layers are oriented in the direction of the major load paths.

Boron is a mineral with some amazing structural properties. Of particular interest are its combination of light weight, high tensile strength, extraordinary stiffness and high heat resistance. Boron filaments are approximately six times stronger and stiffer than comparable aluminum filaments. Boron can withstand 3,700° F without melting, whereas aluminum melts at 1,200 °F. Of the most advanced materials used in sophisticated design, neither titanium, beryllium, magnesium, nor special alloy steels possess boron's favourable combination of characteristics. With the exception of beryllium all of these materials are comparatively low in stiffness. Beryllium is stiff but not strong. Glass filaments are strong but not stiff. As compared to other materials boron yields a high strength to weight ratio.

Boron fibers are produced by vapor-deposition methods. Heated tungsten wire, about 0.0005 in. in diameter, is moved through a chamber in which boron is deposited on the wire, from gases containing a boron compound. The final diameter of the fiber is about 0.004 in. The fibers are preimpregnated with epoxy and backed up with glass fiber cloth to make tape. A cross section of a laminate consisting of a stack of tapes is shown in Fig. 1.1.

Graphite fibers vary in the order of five to ten microns in diameter. The fibers are bundled with 12,000 – 14,000 fibers in each bundle, and then embedded in epoxy. Graphite fibers have very high tensile strength (~250,000 psi) and very high moduli (order of 40,000,000 psi).

In fiber-reinforced metal composites a metal matrix, usually aluminum, replaces the epoxy in resin matrix composites. Aluminum metal composites are formed by diffusion bonding alternate layers of thin (0.008 in.) aluminum foils and collimated layers of high modulus filaments, see Fig. 1.2.

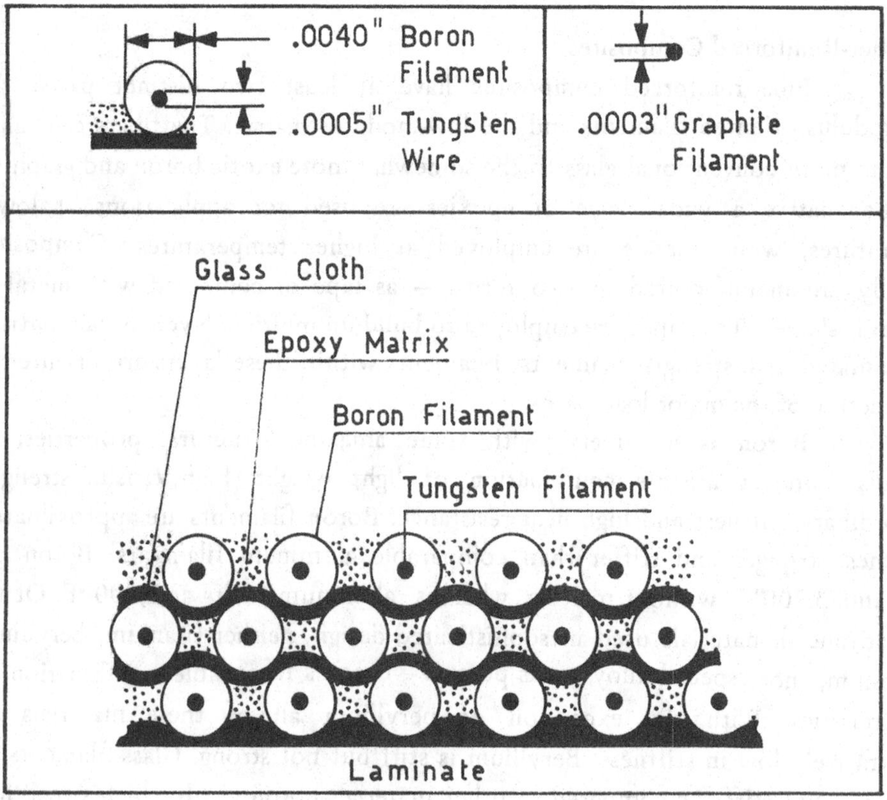

Fig. 1.1. Boron-exposy composite

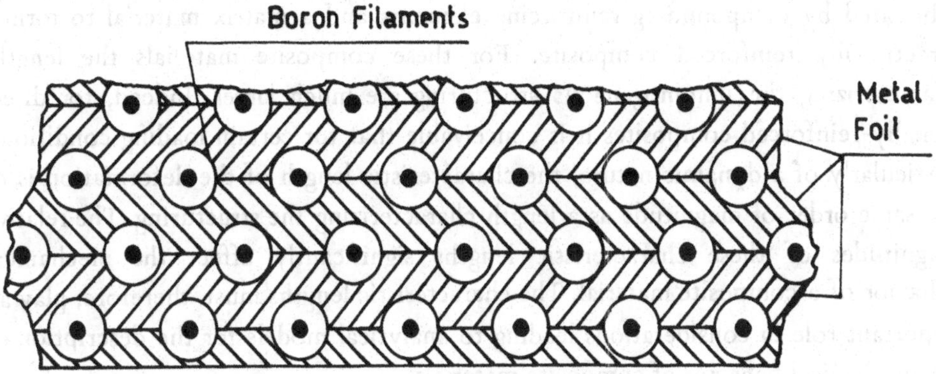

Fig. 1.2. Boron-metal composite

At the present time the relatively high production costs of fiber--reinforced composites and difficulties with the fabrication of structural elements impede a more extensive range of applications. It is, however, widely believed that composite materials will soon emerge as structural materials of low cost and high performance.

1.2. Mathematical Modeling of Fiber-Reinforced Composites

A widespread application of fiber-reinforced composites will require a detailed and reliable knowledge of their physical properties. Because of the potentially great structural and physical variety of fiber-reinforced composites it is not practical to rely solely on experimental determination of these properties. Typical parameters are volume fractions of reinforcing elements and matrix, direction of reinforcements, shapes of fiber cross-sections, and relative fiber positions. Additional variety stems from the physical properties of the constituents. Altogether the variation of the geometrical and physical parameters can lead to an enormous number of possibilities. It is, therefore,clearly desirable to have theories to describe the physical behavior of the composite material in terms of the known geometrical layout of the composite and the known physical properties of the constituents. It will of course be necessary to check such theories by a number of well conceived experiments.

All materials that are used in engineering applications are inhomogeneous. For conventional materials such as metal alloys the characteristic lengths of the inhomogeneities usually are, however, very small as compared to lenghts characterizing deformations. This is not necessarily the case for materials that are fabricated by compounding reinforcing elements and a matrix material to form a directionally reinforced composite. For these composite materials the lengths characterizing the inhomogeneous structuring are much larger. Indeed, for directionally reinforced composites it is conceivable that for certain loading conditions, particularly of a dynamic nature, the characteristic length of the deformation is of the same order of magnitude as a length characterizing the structuring. The relative magnitudes of these characteristic lengths significantly affect the mechanical behavior of a composite material. The characteristic lengths must, therefore, play an important role in considerations leading to analytical models for the description of the mechanical behavior of composite materials.

The mechanical response of a composite can adequately be analyzed on the basis of a theory which accounts for the gross mechanical behavior of the composite material if the ratio of the largest characteristic length of the structuring and the smallest length characterizing the deformation is sufficiently smaller than unity. The simplest aspects of the gross mechanical behavior are described by averages of the displacements, the stresses and the strains over representative elements. In the simplest theory for such averaged quantities the averaged stresses are related to averaged strains by means of effective elastic constants. In this theory, which is termed the "effective modulus theory," the mechanical response of the composite is equivalent to the response of a homogeneous but generally anisotropic medium whose "effective moduli" are determined in terms of the elastic moduli of the constituents and the parameters describing the geometrical layout of the composite. For a detailed treatment of the effective modulus theory for fiber--reinforced materials we refer to the work by Hashin [1.1]. The effective modulus theory is briefly discussed in Section 3.

The advantages of the effective modulus theory, and of more advanced theories as well, are that the actual discrete character of the composite medium is represented within the framework of a homogeneous continuum. Thus, instead of having to deal with numerous sets of field equations (one set for each inhomogeneity), the approximate theory permits us to work with a single set of equations for the composite medium as a whole. For most loading conditions the effective modulus theory is entirely satisfactory. The effective modulus theory becomes, however, less useful for static loading conditions if the interest is focused on local values of the field variables, as for example in the analysis of interface bond failure and other modes of fracture. For dynamic loading conditions the applicability of the effective modulus theory is somewhat restricted even for gross quantities such as frequencies, as will be discussed in the sequel.

1.3. Dynamic Effects

The basic concepts of the motion of elastic continua are reviewed in Section 4. An investigation of the propagation of mechanical disturbances is of importance for bodies that are subjected to high-rate loads such as are generated by impact or by explosive charges. For small times after the application of a high-rate load a pattern of transient waves propagates through the medium. These transient waves may interact rather vigorously with the reinforcing elements. For a directionally reinforced composite the character of the dynamic response depends on the direction of propagation of the disturbances. For wave motions propagating in the direction of the reinforcements, the reinforcing elements act as waveguides. If, on the other hand, the wave motion propagates normal to the direction of the reinforcements the reinforcing elements act essentially as obstacles, which reflect and transmit propagating disturbances. As time increases and the disturbances have been reflected several times from the outer boundaries of the body, a pattern of vibrations sets in.

Mechanical disturbances are subjected to attenuation when propagating in composite materials. The attenuation is effected through geometrical dispersion, and dispersion due to other mechanisms, such as are related to inelastic material behavior, delamination, internal voids and cracks, and crushing of composite constituents. From the point of view of structural integrity, dispersion is considered a desirable phenomenon since it reduces the peak intensity of a stress pulse, and hence its potential to do damage. Of the various dispersion mechanisms geometrical dispersion and dispersion due to inelastic material behavior are best amenable to an analytical treatment.

Geometrical dispersion is the spreading out of pulses caused by interactions at boundaries. The change of shape of a propagating pulse due to geometrical dispersion can be investigated on the basis of an analysis of time-harmonic (sinusoidal) wave trains. For time-harmonic waves, dispersion manifests itself in the dependence of the phase velocity on the wavelength. In a time-harmonic or sinusoidal wave the phase velocity is defined as the rate of advance of points of constant phase angle. An analysis of the propagation of time-harmonic waves results in the derivation of a frequency equation which relates the frequency and the reciprocal wavelength. For linear theories the propagation of a stress pulse can be studied by using a Fourier decomposition of the pulse into modes of time-harmonic wave propagation, see Section 4. Since the frequency equation governs the phase velocities of the individual Fourier components it provides the essential information on geometrical dispersion.

Geometrical dispersion is a well known phenomenon for elastic bodies with elongated boundaries such as rods and layers. On a small scale a directionally reinforced composite essentially is a system of waveguides, and it is thus to be expected that time-harmonic waves, particularly waves that are transverse with respect to the direction of the reinforcing elements, are dispersive while propagating in the composite.

Since the characteristic length of the deformation is expected to be of paramount importance, a first test of approximate theories for the purpose of checking dynamic behavior can be carried out by examining the propagation of time-harmonic waves in a composite medium of infinite extent. Here the characteristic length is the wavelength, which usually enters through the wavenumber $k = 2\pi/\text{wavelength}$. In the presence of dispersion, harmonic waves of different wavelengths propagate at different velocities. This points immediately to a deficiency of the effective modulus theory, because a classical anisotropic continuum model cannot account for the dispersion of free harmonic waves, which should occur in an extended composite medium if the wavelength is of the same order of magnitude as a characteristic length of the structuring. For a layered medium consisting of alternating layers of two elastic materials this restriction was pointed out by Sun, Achenbach and Herrmann [1.2].

For a directionally reinforced composite a new theory is worked out in Part II. This theory improves on the effective modulus theory, in that it can provide a more accurate description of typically dynamic effects. The governing equations are very similar in form to those of a theory of elasticity with microstructure which was developed by Mindlin [1.3].

For a recent article which reviews waves and vibrations in directionally reinforced composites we refer to Ref. [1.4].

CHAPTER 2

CLASSICAL LINEARIZED ELASTICITY

The linearized theory of elasticity has been the subject of several treatises. For a detailed treatment we refer to the book by Sokolnikoff [1.5]. The basic equations are, however, briefly summarized in Sections 2.1–2.3 for the purpose of reference. As a preliminary to the definition of the effective modulus theory some exact solutions for homogeneous elastic bodies of arbitrary shape are given in Section 2.4.

2.1. Notation and Mathematical Preliminaries

A system of fixed rectangular Cartesian coordinates is sufficient for the presentation of the theory. In indicial notation, the coordinate axes may be denoted by x_j and the base vector by $\underset{\sim}{i}_j$, where $j = 1,2,3$. If the components of a vector $\underset{\sim}{u}$ are denoted by u_j, we have

$$\underset{\sim}{u} = u_1 \underset{\sim}{i}_1 + u_2 \underset{\sim}{i}_2 + u_3 \underset{\sim}{i}_3 . \qquad (2.1)$$

Since summations of the type (2.1) frequently occur, we introduce the summation convention, whereby a repeated subscript implies a summation. Equation (2.1) may then be rewritten as

$$\underset{\sim}{u} = u_j \underset{\sim}{i}_j \qquad (2.2)$$

As another example of the use of the summation convention, the scalar product of two vectors is expressed as

$$\underset{\sim}{u} \cdot \underset{\sim}{v} = u_j v_j = u_1 v_1 + u_2 v_2 + u_3 v_3 . \qquad (2.3)$$

As opposed to the free index in u_j, which may assume any one of the values 1,2,3, the index j in (2.2) and (2.3) is a bound index or a dummy index, which must assume all three values 1, 2 and 3.

Quantities with two free indices as subscripts, such as τ_{ij}, denote components of a tensor of second rank $\underset{\sim}{\tau}$, and similarly three free indices define a tensor of rank three. A well-known special tensor of rank two is the Kronecker delta, whose components are defined as

$$(2.4) \qquad \delta_{ij} = \begin{cases} 1 & \text{if } i = j \\ 0 & \text{if } i \neq j. \end{cases}$$

A frequently-used special tensor of rank three is the alternating tensor, whose components are defined as follows:

$$(2.5) \qquad e_{ijk} = \begin{cases} +1 & \text{if ijk represents an even permutation of 123} \\ 0 & \text{if any two of the ijk indices are equal} \\ -1 & \text{if ijk represents an odd permutation of 123.} \end{cases}$$

By use of the alternating tensor and the summation convention, the components of the cross product $\underset{\sim}{h} = \underset{\sim}{u} \wedge \underset{\sim}{v}$ may be expressed as

$$(2.6) \qquad h_i = e_{ijk} \, u_j \, v_k \, .$$

In a vector field, denoted by $\underset{\sim}{u}(\underset{\sim}{x},t)$, the components are denoted by $u_i\,(x_1,x_2,x_3,t)$. Assuming that the functions $u_i\,(x_1,x_2,x_3,t)$ are differentiable, the nine partial derivatives $\partial u_i\,(x_1,x_2,x_3,t)\,/\,\partial x_j$ can be written in indicial notation as $u_{i,j}$. It can be shown that $u_{i,j}$ are the components of a second-rank tensor.

Gauss' theorem

This theorem relates a volume integral to a surface integral over the bounding surface of the volume. Consider a convex region B of volume V, bounded by a surface S which possesses a piecewise continuously turning tangent plane. Such a region is said to be regular. Now let us consider a tensor field $\tau_{jk\ell\ldots p}$, and let every component of $\tau_{jk\ell\ldots p}$ be continuously differentiable in B. Then Gauss' theorem states

$$(2.7) \qquad \int_V \tau_{jk\ell\ldots p,i}\, dV = \int_S n_i\, \tau_{jk\ell\ldots p}\, dA \, ,$$

where n_i are the components of the unit vector along the outer normal to the surface S. If Eq. (2.7) is written with the three components of a vector $\underset{\sim}{u}$ successively substituted for $\tau_{jk\ell\ldots p}$, and if the three resulting equations are added, the result is

$$\int_V u_{i,i}\, dV = \int_S n_i u_i \, dA \, . \tag{2.8}$$

Equation (2.8) is the well-known divergence theorem of vector calculus which states that the integral of the outer normal component of a vector over a closed surface is equal to the integral of the divergence of the vector over the volume bounded by the closed surface.

Notation

The equations governing the linearized theory of elasticity are presented in the following commonly used notation:

position vector:	$\underset{\sim}{x}$	(coordinates x_i)	(2.9)
displacement vector:	$\underset{\sim}{u}$	(components u_i)	(2.10)
small strain tensor:	$\underset{\sim}{\epsilon}$	(components ϵ_{ij})	(2.11)
stress tensor:	$\underset{\sim}{\tau}$	(components τ_{ij})	(2.12)

2.2. Kinematics and Dynamics

Deformation

Let the field defining the displacements of particles be denoted by $\underset{\sim}{u}(\underset{\sim}{x},t)$. As a direct implication of the notion of a continuum, the deformation of the medium can be expressed in terms of the gradients of the displacement vector. Within the restrictions of the linearized theory the deformation is described in a very simple manner by the small-strain tensor $\underset{\sim}{\epsilon}$, with components

$$\epsilon_{ij} = \frac{1}{2}\,(u_{i,j} + u_{j,i}) \, . \tag{2.13}$$

It is evident that $\epsilon_{ij} = \epsilon_{ji}$, i.e., $\underset{\sim}{\epsilon}$ is a symmetric tensor of rank two. It is also useful to introduce the rotation tensor $\underset{\sim}{\omega}$, whose components are defined as

$$\omega_{ij} = \frac{1}{2}\,(u_{i,j} - u_{j,i}) \, . \tag{2.14}$$

We note that ω is an antisymmetric tensor, $\omega_{ij} = -\,\omega_{ji}$.

Linear momentum and the stress tensor

A basic postulate in the theory of continuous media is that the mechanical action of the material points which are situated on one side of an arbitrary material surface within a body upon those on the other side can be completely accounted for by prescribing a suitable surface traction on this surface. Thus if a surface element has a unit outward normal $\underset{\sim}{n}$ we introduce the surface tractions $\underset{\sim}{t}$, defining a force per unit area. The surface tractions generally depend on the orientation of $\underset{\sim}{n}$ as well as on the location $\underset{\sim}{x}$ of the surface element. By means of the "tetrahedron argument" it can easily be shown that the components of the surface traction are related to the components of the stress tensor by

$$t_\ell = \tau_{k\ell} n_k \; . \tag{2.15}$$

Equation (2.15) is the Cauchy stress formula. Physically $\tau_{k\ell}$ is the component in the x_ℓ -direction of the traction on the surface with the unit normal $\underset{\sim}{i}_k$.

According to the principle of balance of linear momentum, the instantaneous rate of change of the linear momentum of a body is equal to the resultant external force acting on the body at the particular instant of time. In the linearized theory this leads to the equation

$$\tau_{k\ell,k} + \rho f_\ell = \rho \ddot{u}_\ell , \tag{2.16}$$

where f_ℓ are the components of body forces. This is Cauchy's first law of motion.

Balance of moment of momentum

For the linearized theory the principle of moment of momentum leads to the result

$$e_{k\ell m} \tau_{\ell m} = 0 \tag{2.17}$$

This result implies that

$$\tau_{\ell m} = \tau_{m\ell} , \tag{2.18}$$

i.e., the stress tensor is symmetric.

2.3. The Homogeneous, Isotropic, Linearly Elastic Solid

In general terms, the linear relation between the components of the stress tensor and the components of the strain tensor is

$$\tau_{ij} = C_{ijk\ell} \, \epsilon_{k\ell} ,$$

It can be shown that elastic isotropy implies that the constants $C_{ijk\ell}$ may be expressed as

$$C_{ijk\ell} = \lambda \delta_{ij} \delta_{k\ell} + \mu (\delta_{ik} \delta_{j\ell} + \delta_{i\ell} \delta_{jk}) .$$

Hooke's law then assumes the well-known form

$$\tau_{ij} = \lambda \epsilon_{kk} \delta_{ij} + 2 \mu \epsilon_{ij} \tag{2.19}$$

Equation (2.19) contains two elastic constants λ and μ, which are known as Lamé's elastic constants.

Putting $j = i$ in Eq. (2.19), thus implying a summation, we obtain

$$\tau_{ii} = (3\lambda + 2\mu)\epsilon_{ii} , \tag{2.20}$$

where we have used $\delta_{ii} = \delta_{11} + \delta_{22} + \delta_{33} = 3$

Stress and strain deviators

The stress tensor can be written as the sum of two tensors, one representing a spherical or hydrostatic stress in which each normal stress component is $1/3\ \tau_{kk}$ and all shear stresses vanish. The complementary tensor is called the stress deviator, denoted by s_{ij}. Thus, the components of the stress deviator are defined by

$$s_{ij} = \tau_{ij} - \frac{1}{3} \tau_{kk} \delta_{ij} . \tag{2.21}$$

In the same manner we can define the strain deviator e_{ij} by

$$e_{ij} = \epsilon_{ij} - \frac{1}{3} \epsilon_{kk} \delta_{ij} . \tag{2.22}$$

From Eq. (2.19) it can now quite easily be shown that the following simple relation exists between s_{ij} and e_{ij}:

$$s_{ij} = 2\mu\, e_{ij} , \tag{2.23}$$

where μ is the shear modulus. In addition we also have, according to (2.20),

$$\frac{1}{3} \tau_{kk} = B\, \epsilon_{kk} ,$$

where

$$B = \lambda + \frac{2}{3}\mu \tag{2.25}$$

is the bulk modulus. Equations (2.24) and (2.25) are completely equivalent to Hooke's law, and these equations may thus also be considered as the constitutive equations for a homogeneous, isotropic, linearly elastic solid.

Strain energy

By the definition of the strain energy density U, we have

$$dU = \tau_{ij}\, d\epsilon_{ij} .$$

In terms of the stress and strain deviators dU assumes the form

$$dU = \left(s_{ij} + \frac{1}{3}\tau_{kk}\delta_{ij}\right) d\left(e_{ij} + \frac{1}{3}\epsilon_{kk}\delta_{ij}\right) ,$$

which reduces to

$$dU = \frac{1}{3}\tau_{kk}\, d\epsilon_{kk} + s_{ij}\, de_{ij} .$$

Use of Eqs. (2.23) and (2.24) then leads to the integrated form

$$U = \frac{1}{2} B\, (\epsilon_{kk})^2 + \mu e_{ij}\, e_{ij} , \tag{2.26}$$

where it is assumed that U vanishes in the undeformed reference state. The isotropic strain energy density function can be written in the alternative form

$$U = \frac{1}{2}\lambda(\epsilon_{kk})^2 + \mu\,\epsilon_{ij}\,\epsilon_{ij} . \tag{2.27}$$

2.4. Exact Solutions for Homogeneous Bodies of Arbitrary Shape

In this section we shall consider two types of boundary conditions for which the quasi-static elasticity problem can be solved for homogeneous bodies of arbitrary shape. For quasi-static equilibrium, and in the absence of body forces, the stress-equations of motion (2.16) reduce to

$$\tau_{ij,j} = 0 \tag{2.28}$$

Substitution of the generalized Hooke's law yields

$$C_{ijk\ell}\, u_{k,\ell j} = 0 \tag{2.29}$$

In the first case linear boundary displacements are applied to the entire surface S, i.e.,

$$u_i\,(S) = \epsilon^o_{ij}\, x_j \quad , \tag{2.30}$$

where ϵ^o_{ij} are symmetric constants and x_j are the surface coordinates. The corresponding displacement field inside the body is now given by

$$u_i\,(x) = \epsilon^o_{ij}\, x_j \tag{2.31}$$

It is obvious that this solution satisfies the prescribed boundary condition (2.30) and the governing system of partial differential equations (2.29).

If the displacement field (2.31) is substituted into the strain-displacement relations we find

$$\epsilon_{ij}\,(\underset{\sim}{x}) = \epsilon^o_{ij} \quad , \tag{2.32}$$

The corresponding stresses are

$$\tau_{ij} = C_{ijk\ell}\, \epsilon^o_{k\ell} \tag{2.33}$$

Thus, the components of the stress tensor are constants.

In the second case tractions of the forms

$$t_i\,(S) = \tau^o_{ij}\, n_j \tag{2.34}$$

are prescribed. Here τ^o_{ij} are symmetric constants and n_j are the components of the outward normal to the surface. The solution to this problem is

$$u_i\,(x) = \alpha_{ij}\, x_j \quad , \tag{2.35}$$

where the constants α_{ij} are defined by

$$\alpha_{ij} = S_{ijk\ell}\, \tau^o_{k\ell} \tag{2.36}$$

In Eq. (2.36) $S_{ijk\ell}$ are the elastic compliances, i.e.,

$$\epsilon_{ij} = S_{ijk\ell}\, \tau_{k\ell} \tag{2.37}$$

and

$$S_{ijrs}\, C_{rsk\ell} = \frac{1}{2}\,(\delta_{ik}\delta_{j\ell} + \delta_{i\ell}\,\delta_{jk}) \tag{2.38}$$

Clearly (2.35) satisfies the equation of equilibrium (2.29). Substitution of (2.35) into

$$t_i = \tau_{ij}\; n_j = C_{ijk\ell}\; u_{k,\ell}\, n_j \tag{2.39}$$

yields

$$t_i = C_{ijk\ell}\, \alpha_{km}\, \delta_{m\ell}\, n_j = C_{ijk\ell}\, S_{k\ell mn}\, \tau^{o}_{mn}\, n_j \tag{2.40}$$

This expression reduces to (2.34) by virtue of Eq. (2.38). It is also noted that the strains are constants, and are given by (2.36). The components of the stress tensor are

$$\tau_{ij}\; (x) = \tau^{o}_{ij} \tag{2.41}$$

The conclusion of this section is that boundary conditions of the types (2.30) and (2.34), applied to homogeneous elastic bodies, yield uniform fields of strain and stress. These simple results do not hold for heterogeneous bodies. As we shall see in the sequel, the boundary conditions (2.30) and (2.34) do, however, play an important role in the definition of the effective mechanical behavior of heterogeneous bodies.

CHAPTER 3

THE EFFECTIVE MODULUS THEORY

In the effective modulus theory the mechanical behavior of a composite material is represented by a homogeneous but generally anisotropic medium. The effective modulus theory, which has been developed to an advanced level of quantitative analysis, is discussed in some detail in a monograph by Hashin [1.1].

A precise definition of the concept of effective elastic moduli, and means of computing the effective moduli, are given in subsequent sections. Here we will, therefore, limit ourselves to the observation that this theory relates volume averages of the components of the stress tensor (denoted by $\bar{\tau}_{ij}$) to volume averages of the components of the strain tensor (denoted by $\bar{\epsilon}_{k\ell}$) by a general anisotropic linear stress-strain relation of the form

$$\bar{\tau}_{ij} = C^*_{ijk\ell}\,\bar{\epsilon}_{k\ell} \qquad (3.1)$$

In Eq. (3.1) we have employed the usual indicial notation with respect to a cartesian coordinate system x_i, $i = 1,2,3$. The elastic constants $C^*_{ijk\ell}$, which are called the effective elastic moduli, are expressions in terms of the material constants of the constituents and the parameters defining the geometrical layout of the composite.

3.1. Theorems for Averaged Stresses and Strains

Suppose we consider a two-phase composite body, with phases occupying regions B_f and B_m. The displacement fields in the phases are $u_i^{(f)}(\underset{\sim}{x},t)$ and $u_i^{(m)}(\underset{\sim}{x},t)$, respectively, where t is the time. The total volume of the body is V, the two phases occupy volumes V_f and V_m. The bounding surface of the total body is S, while the interfaces are denoted by S_{fm}.

The volume average, τ_{ij} (t), of $\epsilon_{ij}(\underset{\sim}{x},t)$ is given by

$$\bar{\epsilon}_{ij}(t) = \frac{1}{V}\int_V \epsilon_{ij}(\underset{\sim}{x},t)\,dV \qquad (3.2)$$

The volume average of the stress is similarly defined as

$$\bar{\tau}_{ij}(t) = \frac{1}{V}\int_V \tau_{ij}(\underset{\sim}{x},t)\,dV \tag{3.3}$$

Suppose the displacements are prescribed on S, i.e.,

$$u_i(S,t) = u_i^o \tag{3.4}$$

and

$$u_i^{(f)} = u_i^{(m)} \quad \text{on} \quad S_{fm} \tag{3.5}$$

then

$$\bar{\epsilon}_{ij}(t) = \frac{1}{2V}\int_S (u_i^o n_j + u_j^o n_i)\,dS \tag{3.6}$$

To prove Eq. (3.6) the strain-displacement relations are substituted in Eq. (3.2), to yield

$$2\bar{\epsilon}_{ij}(t) = \frac{1}{V}\left[\int_{V_f}\left(u_{i,j}^{(f)} + u_{j,i}^{(f)}\right)dV + \int_{V_m}\left(u_{i,j}^{(m)} + u_{j,i}^{(m)}\right)dV\right] \tag{3.7}$$

By using Gauss' theorem, Eq. (2.7), we obtain

$$2\bar{\epsilon}_{ij}(t) = \frac{1}{V}\left[\int_{S_f}\left(u_i^{(f)}n_j + u_j^{(f)}n_i\right)dA + \int_{S_m}\left(u_i^{(m)}n_j + u_j^{(m)}n_i\right)dA\right], \tag{3.8}$$

where S_f and S_m are the bounding surfaces of phases B_f and B_m, respectively. The surfaces S_f and S_m contain the interfaces S_{fm} and the external surface S. Since the contributions from S_{fm} to the two integrals in Eq. (3.8) cancel each other, Eq. (3.8) immediately yields the desired result (3.6).

An additional important result follows if

$$u_i(S,t) = \epsilon_{ij}^o(t)x_j \tag{3.9}$$

Then we have

$$\bar{\epsilon}_{ij}(t) = \epsilon_{ij}^o(t) \tag{3.10}$$

This result follows by substituting Eq. (3.9) into (3.6), and observing that

$$\int_S x_k n_j \, dA = \int_V x_{k,j} dV = V\delta_{kj}$$

Next we consider a theorem for the averaged stresses. Suppose on the external surface S the tractions are prescribed as

$$t_i\,(S,t) = \tau_{ij} n_j = t_i^o \tag{3.11}$$

At the interfaces the tractions are continuous

$$t_i^{(f)} = \tau_{ij}^{(f)} n_j = - t_i^{(m)} = - \tau_{ij}^{(m)} n_j \quad \text{on } S_{fm},$$

where it should be understood that n_j are the components of the outward normal. The average stress theorem now states that

$$\bar{\tau}_{ij}(t) = \frac{1}{V}\left[\int_S x_j t_i \, dA + \int_V x_j f_i \, dV\right], \tag{3.12}$$

where f_i are the components of body forces. To prove Eq. (3.12) we first employ the stress-equation of equilibrium

$$\tau_{ij,j} + f_i = 0 \tag{3.13}$$

to verify that

$$\tau_{ij} = (\tau_{ik} x_j)_{,k} + f_i x_j \tag{3.14}$$

Substituting this relation into Eq. (3.3), and employing Gauss' theorem, Eq. (2.7), we find

$$\bar{\tau}_{ij}(t) = \frac{1}{V}\left[\int_{S_f} x_j \tau_{ik}^{(f)} n_k^{(f)} dA + \int_{S_m} x_j \tau_{ik}^{(m)} n_k^{(m)} dA + \int_V f_i x_j \, dV\right] \tag{3.15}$$

Since the tractions are continuous at the interfaces S_{fm}, Eq. (3.15) reduces to (3.12). By virtue of the symmetry of the averaged stress tensor, Eq. (3.12) can be symmetrized to the form

$$\bar{\tau}_{ij}(t) = \frac{1}{2V}\left[\int_S (x_j t_i + x_i t_j) dA + \int_V (x_j f_i + x_i f_j) dV\right] \tag{3.16}$$

An important special result follows if

(3.17) $$t_i\,(S,t) = \tau^{o}_{ij}(t)n_j$$

and

$$f_i\,(\underset{\sim}{x},t) = 0$$

Then we have

(3.18) $$\bar{\tau}_{ij}\,(t) = \tau^{o}_{ij}\,(t).$$

The proof is left to the reader.

3.2. Effective Elastic Moduli

In this section we discuss the concept of effective elastic constants. Effective elastic constants are defined by examining the mechanical response of a representative volume of the composite. The latter is a specimen that (a) is structurally entirely typical of the whole composite on average, and (b) contains a sufficient number of inclusions for the apparent overall moduli to be effectively independent of the surface values of traction and displacement, so long as these values are "macroscopically uniform". That is, they fluctuate about a mean with a wavelength small compared with the dimensions of the volume, and the effects of such fluctuations become insignificant within a few wavelengths of the surface. The contribution of this surface layer to any average can be made negligible by taking the representative volume large enough. The above description of a representative volume was given by Hill [1.5]. Generally the representative volume is taken as a cube.

There are various ways of defining the effective elastic constants. We may consider a representative volume of the medium and subject it to prescribed surface displacements of the kind that would produce a uniform strain ϵ^{o}_{ij} in a homogeneous material. By Eq. (3.10) this is, however, also the volume-averaged strain $\bar{\epsilon}_{ij}$ in a heterogeneous body. A computation of the volume-averaged stress $\bar{\tau}_{ij}$ then yields the effective elastic constants. Alternatively we may subject the body to surface tractions of the kind that would produce a uniform stress τ^{o}_{ij} in a homogeneous material. As a third alternative, displacements may be prescribed over part of the external surface, and tractions over the remaining part.

Let us first consider the case that a representative volume of a composite material is subjected to the boundary condition

$$u_i(S) = \epsilon^o_{ij}\, x_j \tag{3.19}$$

At least conceptually the problem of determining the corresponding distribution of the stresses and the displacements can be solved by established methods of the linearized theory of elasticity. In this section the purpose is to establish a relation between volume averages.

It is not difficult to see that the complete displacement distribution inside the medium can be written as the superposition of the displacements due to six separate boundary displacement vectors. Suppose the displacement due to $\epsilon^o_{k\ell} = 1$ is denoted by $u_i^{(k\ell)}(\underset{\sim}{x})$, then by linear superposition

$$u_i(\underset{\sim}{x}) = \epsilon^o_{k\ell}\, u_i^{(k\ell)}(\underset{\sim}{x}) \quad ,$$

where a summation should be carried out over k and ℓ. The corresponding strain at every point is given by

$$\epsilon_{ij}(\underset{\sim}{x}) = \epsilon^o_{k\ell}\, \epsilon^{(k\ell)}_{ij}(\underset{\sim}{x}) \quad ,$$

where

$$\epsilon^{(k\ell)}_{ij}(\underset{\sim}{x}) = \frac{1}{2}\left(u^{(k\ell)}_{i,j} + u^{(k\ell)}_{j,i}\right)$$

The stress at every point becomes

$$\tau_{ij}(\underset{\sim}{x}) = \epsilon^o_{k\ell}\, C_{ijmn}(\underset{\sim}{x})\, \epsilon^{(k\ell)}_{mn}(\underset{\sim}{x}) \tag{3.20}$$

Here the $C_{ijmn}(\underset{\sim}{x})$ are the space dependent elastic moduli of the heterogeneous material. For a two-phase material these elastic moduli can assume the values $C^{(f)}_{ijmn}$ and $C^{(m)}_{ijmn}$.

Now let us consider the volume averages over the representative volume of the relation (3.20). We find

$$\bar{\tau}_{ij} = C^*_{ijk\ell}\, \epsilon^o_{k\ell} \tag{3.21}$$

where

$$C^*_{ijk\ell} = \frac{1}{V}\int_V C_{ijmn}(\underset{\sim}{x})\, \epsilon^{(k\ell)}_{mn}(\underset{\sim}{x})\, dV \tag{3.22}$$

In view of the result (3.10), Eq. (3.20) can be expressed in the form

$$\bar{\tau}_{ij} = C^*_{ijk\ell}\,\bar{\epsilon}_{k\ell} \tag{3.23}$$

Thus the averaged stress is related to the averaged strain by constants which are defined by Eq. (3.22). The thus-defined elastic constants are called effective moduli, and the corresponding framework of equations defines the effective modulus theory.

By subjecting a composite body to traction boundary conditions of the form

$$t_i\,(S) = \tau^o_{ij}\,n_j\,, \tag{3.24}$$

we can obtain the relation

$$\bar{\epsilon}_{ij} = S^*_{ijk\ell}\,\bar{\tau}_{k\ell} \tag{3.25}$$

The constants $S^*_{ijk\ell}$ are called the effective compliances.

By the quotient law of tensor analysis it follows that $C^*_{ijk\ell}$ are tensors of the fourth rank. It is also not difficult to show that the effective elastic constants satisfy the usual symmetry relations, so that there is at most a total of 21 independent effective moduli.

Apparently the first contribution to the computation of effective elastic moduli is due to Voigt [1.7], who considered a polycrystalline aggregate. Voigt assumed that for boundary conditions of the type (3.19) the strain is uniform throughout the representative volume. Equation (3.22) then reduces to

$$C^*_{ijk\ell} = c_f\,C^{(f)}_{ijk\ell} + c_m\,C^{(m)}_{ijk\ell} \tag{3.26}$$

Here c_f and c_m are volume ratios, i.e.,

$$c_f = V_f/V \quad \text{and} \quad c_m = V_m/V \tag{3.27}$$

It follows from Eq. (3.26) that for isotropic constituent materials, the gross behavior then is also isotropic. A useful equivalent form of Eq. (3.24) is then obtained by splitting $\bar{\tau}_{ij}$ and $\bar{\epsilon}_{ij}$ into isotropic and deviatoric parts, see Section 2.3, i.e.,

$$\frac{1}{3}\,\bar{\tau}_{kk} = B^*\,\bar{\epsilon}_{kk}$$

$$\bar{s}_{ij} = 2\mu^*\,e_{ij}$$

where B* and μ^* are the effective bulk modulus and the effective shear modulus, respectively. For the Voigt approximation we then find

$$B_V^* = c_f B^{(f)} + c_m B^{(m)}$$

$$\mu_V^* = c_f \mu^{(f)} + c_m \mu^{(m)}$$

The dual assumption, due to Reuss, [1.8], and also originally proposed for a polycristal, is that the stress is uniform for boundary conditions of the form (3.24). The results are

$$1/B_R^* = c_f / B^{(f)} + c_m / B^{(m)}$$

$$1/\mu_R^* = c_f / \mu^{(f)} + c_m / \mu^{(m)}$$

Neither the Voigt assumption nor the Reuss assumption is correct. For the Voigt assumption the tractions at the phase interfaces are not in equilibrium, while the Reuss assumption implies lack of displacement continuity at the interfaces. The difference between the estimates can be written as

$$B_V^* - B_R^* = \left[B^{(f)} - B^{(m)}\right]^2 \Big/ \left[\frac{B^{(f)}}{c_f} + \frac{B^{(m)}}{c_m}\right]$$

where $c_f + c_m = 1$ has been used. Thus, the Voigt estimate always exceeds the Reuss estimate.

3.3. Computation of the Effective Moduli

The number of independent elastic constants is generally much less than 21, due to the existence of symmetries in the structuring of the material. A plane is called a plane of elastic symmetry if reflection in the plane leaves the stress-strain law unchanged. If rotation of a coordinate direction about an axis leaves elastic properties unchanged, then the axis is an axis of symmetry. In a composite material the symmetry may be small scale, i.e., for the elastic properties at a point, or it may be a gross property due to the structuring of the composite. Here we consider the case that the constituents of the composite are isotropic, i.e., every axis is an axis of symmetry as far as the individual constituents are concerned, and anisotropy is in evidence only for the gross properties of the medium.

For a fiber-reinforced composite the significant cases with regard to gross anisotropy are orthotropy, square symmetry and transverse isotropy. An orthotropic elastic body has three mutually perpendicular planes of elastic symmetry. An example of such a material is a fiber-reinforced composite with a rectangular array of identical circular fibers, see Fig. 1.2a. A square-symmetric material is an orthotropic material in which the two axis x_2 and x_3 are elastically equivalent. This means that the stress-strain law is insensitive to a 90° rotation around the x_1 axis. The square array of fibers shown in Fig. 1.2b gives rise to a square-symmetric material. For a transversely isotropic material one axis, say the x_2-axis, is an axis of rotational symmetry. Thus, the constitutive relations are insensitive to any rotation around the x_2-axis. The hexagonal array of identical fibers shown in Fig. 1.2c gives rise to transversely isotropic gross properties. Transverse isotropy may also be assumed for a random distribution of unidirectional fibers across the cross-section. Another example of gross transverse isotropy is provided by a laminated medium consisting of alternating layers of two isotropic materials, as shown in Fig. 1.2d.

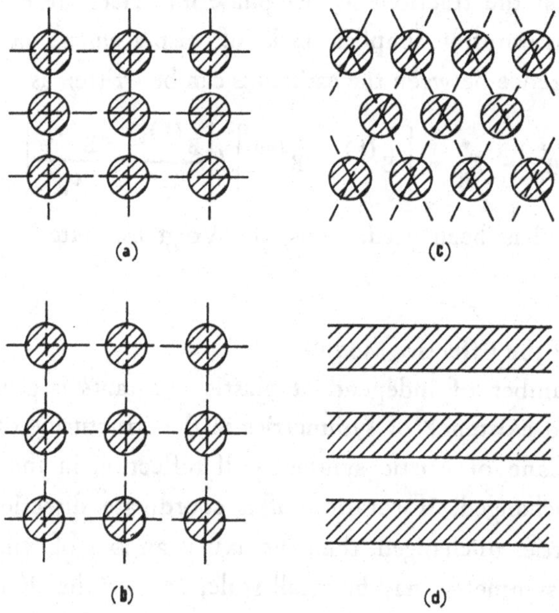

Fig. 1.2. Directionally reinforced composites: (a) fiber-reinforced composite-rectangular array, (b) square array, (c) hexagonal array, (d) laminated medium.

Let us consider the case of transverse isotropy in somewhat more detail. A transversely isotropic material is defined by five elastic constants. Choosing the x_2-axis as the axis of symmetry, the effective stress-strain relations are

$$\bar{\tau}_{11} = C^*_{11}\,\bar{\epsilon}_{11} + C^*_{12}\,\bar{\epsilon}_{22} + C^*_{13}\,\bar{\epsilon}_{33} \tag{3.28}$$

$$\bar{\tau}_{22} = C^*_{12}\,\bar{\epsilon}_{11} + C^*_{22}\,\bar{\epsilon}_{22} + C^*_{12}\,\bar{\epsilon}_{33} \tag{3.29}$$

$$\bar{\tau}_{33} = C^*_{13}\,\bar{\epsilon}_{11} + C^*_{12}\,\bar{\epsilon}_{22} + C^*_{11}\,\bar{\epsilon}_{33} \tag{3.30}$$

$$\bar{\tau}_{23} = 2C^*_{44}\,\bar{\epsilon}_{23} \tag{3.31}$$

$$\bar{\tau}_{13} = (C^*_{11} - C^*_{13})\,\bar{\epsilon}_{13} \tag{3.32}$$

$$\bar{\tau}_{12} = 2C^*_{44}\,\bar{\epsilon}_{12} \tag{3.33}$$

Here we have used the simplified notation for elastic constants, which is commonly used for transversely isotropic materials.

We will consider a cylindrical specimen with the generators parallel to the fibers. The x_2-axis (the axis of symmetry), is also parallel to the fibers. Let us examine the boundary conditions that should be applied to compute the effective moduli. For the determination of C^*_{22} and C^*_{12} we apply the boundary conditions

$$u_2(S) = \overset{o}{\epsilon}_{22}\,x_2 \tag{3.34}$$

$$u_1(S) = u_3(S) = 0 \tag{3.35}$$

By virtue of the average strain theorem, Eq. (3.6), the average strains are

$$\bar{\epsilon}_{22} = \overset{o}{\epsilon}_{22}\ , \quad \text{all other } \bar{\epsilon}_{ij} \equiv 0$$

Then from Eqs. (3.28)-(3.30) we find

$$\bar{\tau}_{22} = C^*_{22}\,\overset{o}{\epsilon}_{22}$$
$$\bar{\tau}_{11} = \bar{\tau}_{33} = C^*_{12}\,\overset{o}{\epsilon}_{22}$$

It is noted that Eqs. (3.34) and (3.35) correspond to uniaxial straining with transverse deformation prevented. To determine C^*_{22} and C^*_{12} we must compute the axial and transverse average stresses.

Next we impose the boundary conditions

$$u_1(S) = \epsilon^o x_1 \;,\; u_3(S) = \epsilon^o x_3 \text{ and } u_2(S) = 0$$

The average strains are by virtue of Eq. (3.6)

$$\bar{\epsilon}_{11} = \epsilon^o \text{ and } \bar{\epsilon}_{33} = \epsilon^o \;,\; \text{all other } \bar{\epsilon}_{ij} \equiv 0$$

By substituting these strains into (3.28)-(3.30) we find

$$\bar{\tau}_{22} = 2C^*_{12}\,\epsilon^o$$

$$\tau_{11} = \tau_{33} = (C^*_{11} + C^*_{13})\,\epsilon^o$$

The effective constants $\mu^*_T = (\frac{1}{2})(C^*_{11} - C^*_{13})$ and $\mu^*_A = C^*_{44}$ are evidently effective shear moduli. It is observed that μ^*_T governs shearing in the transverse (x_1x_3)-plane, hence μ^*_T is called the effective transverse shear modulus. The modulus μ^*_A is called the effective axial shear modulus, since it governs shearing in the $(x_2\,x_1)$- and $(x_2\,x_3)$- planes. The appropriate boundary conditions are

For μ^*_T:

$$u_2(S) = 0$$
$$u_1(S) = \epsilon^o_{13}\,x_3$$
$$u_3(S) = \epsilon^o_{13}\,x_1$$

For μ^*_A:

$$\left.\begin{aligned} u_2(S) &= \epsilon^o_{23}\,x_3 \\ u_3(S) &= \epsilon^o_{23}\,x_2 \\ u_1(S) &= 0 \end{aligned}\right\} \text{or} \left\{\begin{aligned} u_2(S) &= \epsilon^o_{21}\,x_1 \\ u_1(S) &= \epsilon^o_{21}\,x_2 \\ u_3(S) &= 0 \end{aligned}\right.$$

From the problem for μ^*_T we find

$$\mu^*_T = \bar{\tau}_{13}/2\epsilon^o_{13}$$

while the problem for μ_A^* yields

$$\mu_A^* = \bar{\tau}_{23}/2\epsilon_{23}^{o} \quad \text{or} \quad \mu_A^* = \bar{\tau}_{21}/2\epsilon_{21}^{o}$$

Thus, the method described here requires the computation of volume-averaged stresses for prescribed displacements on the boundary of a representative volume. It is not clear whether another approach, based for example on traction boundary conditions, would yield the same effective moduli.

3.4. The Laminated Medium

As a specific example of a transversely isotropic medium we will consider a laminated medium consisting of alternating plane, parallel layers of two homogeneous, isotropic. elastic materials. The elastic constants and the thicknesses of the high-modulus reinforcing layers and the low-modulus matrix layers are denoted by λ_f, μ_f, d_f and λ_m, μ_m, d_m, respectively, see Fig. 1.3. According to the effective modulus theory, the gross elastic behavior of the laminated medium is transversely isotropic with the x_2-axis as the axis of symmetry. The stress-strain relations can then be described by equations of the general form (3.28)-(3.33). The effective elastic constants C_{11}^*, etc. have been computed by Postma [1.9].

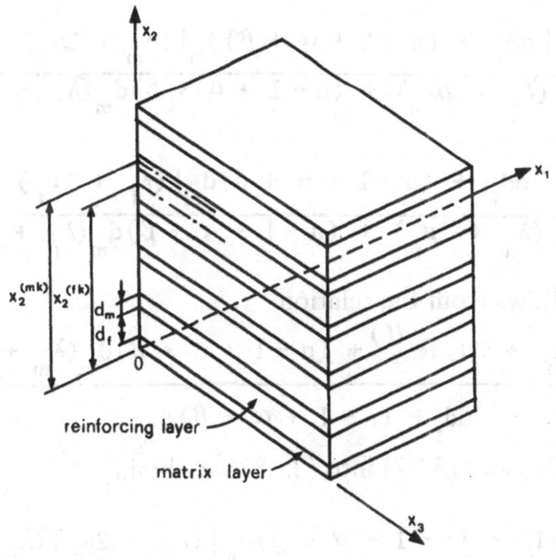

Fig. 1.3. Laminated medium.

An element of the laminated medium is taken to contain n high-modulus layers, n−1 low-modulus layers, and upper and lower strips of low-modulus layers of thicknesses αd_m and βd_m, respectively. We will first compute the constant C^*_{22}. The following boundary conditions are applied to the representative element.

$$\begin{aligned} &\text{on planes } x_1 = \text{constant} : u_1 = 0 \\ &\text{on planes } x_3 = \text{constant} : u_3 = 0 \\ &\text{on planes } x_2 = \text{constant} : u_2 = \epsilon^o_{22}\, x_2 \end{aligned}$$

It then immediately follows that

$$\bar{\epsilon}_{11} = \bar{\epsilon}_{33} \equiv 0 \quad ,$$

while

$$n d_f \epsilon^{(f)}_{22} + (n-1+\alpha+\beta) d_m \epsilon^{(m)}_{22} = [n d_f + (n-1+\alpha+\beta) d_m]\, \epsilon^o_{22}$$

From stress equilibrium it follows that

$$(\lambda_m + 2\mu_m)\, \epsilon^{(m)}_{22} = (\lambda_f + 2\mu_f)\, \epsilon^{(f)}_{22}$$

Solving from these equations for $\epsilon^{(f)}_{22}$ and $\epsilon^{(m)}_{22}$ yields

$$(3.36) \quad \epsilon^{(f)}_{22} = \frac{[n d_f + (n-1+\alpha+\beta) d_m](\lambda_m + 2\mu_m)}{n d_f (\lambda_m + 2\mu_m) + (n-1+\alpha+\beta) d_m (\lambda_f + 2\mu_f)}\, \epsilon^o_{22}$$

$$(3.37) \quad \epsilon^{(m)}_{22} = \frac{[n d_f + (n-1+\alpha+\beta) d_m](\lambda_f + 2\mu_f)}{n d_f (\lambda_m + 2\mu_m) + (n-1+\alpha+\beta) d_m (\lambda_f + 2\mu_f)}\, \epsilon^o_{22}$$

The average stress follows from the relation

$$(3.38) \quad \bar{\tau}_{22} = \frac{n d_f (\lambda_f + 2\mu_f) \epsilon^{(f)}_{22} + (n-1+\alpha+\beta) d_m (\lambda_m + 2\mu_m) \epsilon^{(m)}_{22}}{n d_f + (n-1+\alpha+\beta) d_m}$$

Substituting Eqs. (3.36) and (3.37) into (3.38) we obtain

$$C^*_{22} = \frac{[n d_f + (n-1+\alpha+\beta) d_m](\lambda_f + 2\mu_f)(\lambda_m + 2\mu_m)}{n d_f (\lambda_m + 2\mu_m) + (n-1+\alpha+\beta) d_m (\lambda_f + 2\mu_f)}$$

This expression can, of course, also be obtained by employing the general formula (3.22).

In the limit as $n \to \infty$ this expression simplifies to

$$C^*_{22} = \frac{1}{D}\left[(d_f + d_m)^2 (\lambda_f + 2\mu_f)(\lambda_m + 2\mu_m)\right] \tag{3.40}$$

where

$$D = (d_f + d_m)[d_f(\lambda_m + 2\mu_m) + d_m(\lambda_f + 2\mu_f)]$$

The same result is obtained if the representative volume contains one high-modulus layer and one low-modulus layer.

The other effective moduli can be obtained in a similar manner. The results are

$$C^*_{11} = \frac{1}{D}\left[(d_f + d_m)^2 (\lambda_f + 2\mu_f)(\lambda_m + 2\mu_m) + 4d_f d_m(\mu_f - \mu_m)(\lambda_f + \mu_f - \lambda_m - \mu_m)\right] \tag{3.41}$$

$$C^*_{12} = \frac{1}{D}\,(d_f + d_m)[\lambda_f d_f(\lambda_m + 2\mu_m) + \lambda_m d_m(\lambda_f + 2\mu_f)] \tag{3.42}$$

$$C^*_{13} = \frac{1}{D}\left[(d_f + d_m)^2 \lambda_f \lambda_m + 2(\lambda_f d_f + \lambda_m d_m)(\mu_m d_f + \mu_f d_m)\right] \tag{3.43}$$

$$C^*_{44} = (d_f + d_m)\mu_f \mu_m / (d_f \mu_m + d_m \mu_f) \tag{3.44}$$

3.5. Bounds for the Effective Elastic Constants

The method of computation of the effective elastic moduli which was outlined in Section 3.3 requires the computation of the distributions of stresses and/or strains in the interior of a representative volume which is subjected to certain boundary conditions. Clearly such computations can become rather difficult for more complicated composite geometries. For cases where it would be too laborious to apply the methods of Section 3.3 it is often possible to obtain upper and lower bounds for the effective elastic moduli by variational methods. Such bounding methods are particularly important when the phase geometry is only partially known, because it is then possible to obtain the bounds in terms of whatever information is available.

In this section we will briefly review the two best known extremum principles for small variations of the elastic fields: the principle of minimum potential energy, and the principle of minimum complementary energy. We will also present an elementary application of these two principles to obtain bounds for effective moduli.

The extremum principles of the theory of elasticity are discussed in many textbooks. For a homogeneous body a derivation can, for example be found in the book by Fung, [1.10]. It is not difficult to work out the extension to a multi-phase elastic body whose phases are homogeneous and anisotropic, see Hashin, [1.1].

Suppose that the boundary conditions on the multi-phase elastic body are

(3.45) $$u_i(S) = u_i^o \qquad \text{on } S_u$$

(3.46) $$t_i(S) = C_{ijk\ell} u_{k,\ell} n_j \qquad \text{on } S_t$$

The governing equations in the m-th phase are of the form

(3.47) $$C_{ijk\ell}^{(m)} u_{k,\ell j}^{(m)} = 0 \qquad \text{in } V_m$$

The conditions at the interfaces of the phases are

(3.48) $$u_i \text{ is continuous} \qquad \text{on } S_{int}$$

(3.49) $$t_i = \tau_{ij} n_j = C_{ijk\ell} u_{k,\ell} n_j \text{ is continuous} \qquad \text{on } S_{int}$$

The solution to the problem defined by Eqs. (3.45)-(3.49) is denoted by

$$u_i = u_i^{(m)} \qquad \text{in } V_m$$

By the usual methods the potential energy of the system can now be derived as

(3.50) $$U_P = \sum_m \int_{V_m} W_m^e \, dV - \int_{S_t} t_i^o u_i \, dA$$

Where the summation is carried out over all the phases, and where W_m^e is the strain energy density of the m-th phase, i.e.,

$$W^{\epsilon}_{m} = \frac{1}{2} C^{(m)}_{ijk\ell} u^{(m)}_{i,j} u^{(m)}_{k,\ell} = \frac{1}{2} C^{(m)}_{ijk\ell} \epsilon^{(m)}_{ij} \epsilon^{(m)}_{k\ell} \tag{3.51}$$

Now, let us define an admissible displacement field $\tilde{u}_i$, which is continuous throughout the phases, and also satisfies the following requirements

$$\tilde{u}_i = u^o_i \quad \text{on } S_u$$

$$\tilde{u}_i \text{ is continuous} \quad \text{on } S_{int}$$

and we define the difference field Δu_i by

$$\Delta u_i = \tilde{u}_i - u_i$$

It follows that

$$\Delta u_i = 0 \quad \text{on } S_u$$

$$\Delta u_i \text{ is continuous} \quad \text{on } S_{int}$$

We also define

$$\Delta u_i = \Delta u_i^{(m)} \quad \text{in } V_m$$

and

$$\Delta \epsilon^{(m)}_{ij} = \frac{1}{2}\left(\Delta u^{(m)}_{i,j} + \Delta u^{(m)}_{j,i}\right)$$

The so-called potential energy functional is obtained by replacing u_i in Eqs. (3.50) and (3.51) by $\tilde{u}_i$. We have

$$\tilde{U}_P = \sum_m \int_{V_m} \tilde{W}^{\epsilon}_m dV - \int_{S_t} t^o_i \tilde{u}_i \, dA ,$$

where

$$\tilde{W}^{\epsilon}_m = \frac{1}{2} C^{(m)}_{ijk\ell} \tilde{u}^{(m)}_{i,j} \tilde{u}^{(m)}_{k,\ell}$$

It is now not difficult to show that

$$\tilde{U}_P - U_P = \frac{1}{2} \sum_m \int_{V_m} C^{(m)}_{ijk\ell} \, \Delta\epsilon^{(m)}_{k\ell} \, \Delta\epsilon^{(m)}_{ij} \, dV$$

From the positive definiteness property of strain energy it then follows that

$$\tilde{U}_P \geq U_P \, , \tag{3.52}$$

where the equality sign holds if the admissible displacement happens to be the true elastic displacement. The inequality (3.52) expresses the principle of minimum potential energy: The potential energy is the absolute minimum of the potential energy functional.

For the special case that the displacements are prescribed over the entire surface Eq. (3.52) reduces to the statement: For a body with displacements prescribed over its entire surface the strain energy is the absolute minimum of the strain energy functional.

The principle of minimum complementary energy can be derived in a completely analogous manner. We find

$$\tilde{U}_C \geq U_C \, , \tag{3.53}$$

where the complementary energy, U_C, is defined as

$$U_C = \sum_m \int_{V_m} W^{\tau}_m \, dV - \int_{S_t} t_i \, u^o_i \, dA \, , \tag{3.54}$$

with the following definition for the stress energy density in the m-th phase

$$W^{\tau}_m = \frac{1}{2} S^{(m)}_{ijk\ell} \, \tau^{(m)}_{ij} \, \tau^{(m)}_{k\ell} \tag{3.55}$$

In eq. (3.55) the constants $S^{(m)}_{ijk\ell}$ are the elastic compliances. The complementary energy functional, $\tilde{U}_C$ is obtained by substituting an admissible stress field in Eq. (3.54) of the following form

$$\tau^{(m)}_{ij} = \tilde{\tau}^{(m)}_{ij} \qquad \text{in } V_m$$

which satisfies the following requirements

$$\tilde{\tau}_{ij} n_j = t^o_i \qquad \text{on } S_t$$

$$\tilde{\tau}_{ij,j} = 0 \qquad \text{in } V_m$$

$$\tilde{t}_i = \tau_{ij} n_j = \text{continuous} \qquad \text{on } S_{int}$$

Equation (3.53) states that the complementary energy is the absolute minimum of the complementary energy functional.

For the important special case that tractions are prescribed over the entire surface, i.e., $S_u \equiv 0$, Eq. (3.53) reduces to the statement that the stress energy is the absolute minimum of the stress energy functional.

For an elementary example of the computation of bounds for the effective elastic moduli, by means of extremum principles, we return to the transversely isotropic material discussed in Section 3.3. Suppose we consider the computation of the axial shear modulus μ^*_A. In the displacement formulation, displacements are prescribed over the entire surface, according to, for example,

$$u_2(S) = \epsilon^o_{23} x_3 \tag{3.56}$$

$$u_3(S) = \epsilon^o_{23} x_2 \tag{3.57}$$

$$u_1(S) = 0 \tag{3.58}$$

The boundary conditions for the dual traction formulation are

$$t_1(S) = 0 \tag{3.59}$$

$$t_2(S) = \tau^o_{23} n_3 \tag{3.60}$$

$$t_3(S) = \tau^o_{23} n_2 \tag{3.61}$$

In the formulation given by Eqs. (3.56)-(3.58) displacements are prescribed over the entire surface. In that case the principle of minimum strain energy is pertinent. The boundary conditions suggest the following admissible displacements

$$\tilde{u}_1 = 0, \; \tilde{u}_2 = \epsilon^o_{23} x_3 \; \text{ and } \; \tilde{u}_3 = \epsilon^o_{23} x_2$$

This set of displacements satisfies the boundary conditions, and it is continuous at the phase interfaces. For deformation in axial shear the strain energy is

$$U^{\epsilon} = 2\mu_A^*(\epsilon_{23}^{o})^2 V$$

If the phases are isotropic the strain energy functional is easily computed as

$$\tilde{U}^{\epsilon} = 2\left(\mu^{(f)} c_f + \mu^{(m)} c_m\right)\left(\epsilon_{23}^{o}\right)^2 V$$

By virtue of (3.52) it then follows that

$$\mu_A^* \leq \mu^{(f)} c_f + \mu^{(m)} c_m \tag{3.62}$$

In the formulation (3.59)-(3.61) the tractions are prescribed over the entire surface, and the principle of minimum stress energy is pertinent. An admissible stress field is

$$\tilde{\tau}_{23} = \tau_{23}^{o}\,,\ \tilde{\tau}_{32} = \tau_{23}^{o}\,,\text{all other } \tilde{\tau}_{ij} \equiv 0$$

For axial shear the stress energy is

$$U^{\tau} = (\tau_{23}^{o})^2 V/2\mu_A^*$$

For isotropic phases the stress energy functional becomes

$$\tilde{U}^{\tau} = \frac{1}{2}\left(\tau_{23}^{o}\right)^2\left(c_f/\mu^{(f)} + c_m/\mu^{(m)}\right) V$$

By virtue of Eq. (3.53) we find

$$\mu_A^* \geq \left(c_f/\mu^{(f)} + c_m/\mu^{(m)}\right)^{-1} \tag{3.63}$$

The results (3.62) and (3.63) may be rewritten as

$$\left[c_f/\mu^{(f)} + c_m/\mu^{(m)}\right]^{-1} \leq \mu_A^* \leq c_f\mu^{(f)} + c_m\mu^{(m)}$$

It is interesting to note that the bounds correspond to the Voigt and Reuss approximations, respectively, see Section 3.3.

Bounds for the other effective moduli may be obtained analogously. A discussion of bounds can be found in the article by Hill, see Ref. [1.6].

CHAPTER 4

WAVE MOTIONS IN A LAMINATED MEDIUM

4.1. The Propagation of Mechanical Disturbances

Wave propagation is a classical subject which has been discussed in great generality in several treatises. A detailed discussion of the propagation of mechanical disturbances, in particularly in elastic solids, can be found in a recent book by Achenbach [1.11].

Mechanical waves originate in the forced motion of a portion of a deformable medium. As elements of the medium are deformed the disturbance is transmitted from one point to the next and the disturbance, or wave, progresses through the medium. In this process the resistance offered to deformation by the consistency of the medium, as well as the resistance to motion offered by inertia, must be overcome. As the disturbance propagates it carries along amounts of energy in the forms of kinetic and potential energies. The transmission of energy is effected because motion is passed on from one particle to the next, and not by any sustained bulk motion of the entire medium. Mechanical waves are characterized by the transport of energy through motions of particles about an equilibrium position.

Deformability and inertia are essential properties of a medium for the transmission of mechanical wave motions. If the medium were not deformable any part of the medium would immediately experience a disturbance in the form of an internal force or an acceleration upon application of a localized excitation. Similarly, if a hypothetical medium were without inertia there would be no delay in the displacement of particles and the transmission of the disturbance from particle to particle would be effected instantaneously to the most distant particle. Indeed, it can be shown analytically that the velocity of propagation of a mechanical disturbance always assumes the form of the square root of the ratio of a parameter defining the resistance to deformation and a parameter defining the inertia of the medium. All real materials are of course deformable and possess mass and thus all real materials transmit mechanical waves.

In mathematical terms the simplest traveling wave in one dimension is defined by an expression of the type $f = f(x\text{-}ct)$ where f as a function of the spatial coordinate x and the time t represents a disturbance in the values of some physical quantity. For mechanical waves f generally denotes a displacement, a particle velocity or a stress component. The function $f(x\text{-}ct)$ is called a simple wave function and the argument x-ct is the phase of the wave function. If t is increased by any value, say Δt, and simultaneously x is increased by $c\Delta t$, the value of $f(x\text{-}ct)$ is clearly not altered. The function $f(x\text{-}ct)$ thus represents a disturbance advancing in the positive x-direction with a velocity c. The velocity c is termed the phase velocity. The propagating disturbance represented by $f(x\text{-}ct)$ is a special wave in that the shape of the disturbance is unaltered as it propagates through the medium.

Progressive harmonic waves

A progressive or traveling harmonic wave is represented by an expression of the form

$$u(x,t) = A \cos [k(x - ct)] , \tag{4.1}$$

where the amplitude A is independent of x and t. Equation (4.1) is of the general form $f(x\text{-}ct)$ and thus clearly represents a traveling wave. The argument k(x-ct) is called the phase of the wave: points of constant phase are propagated with the phase velocity c. At any instant t, u(x,t) is a periodic function of x with wavelength Λ, where $\Lambda = 2\pi/k$. The quantity $k = 2\pi/\Lambda$, which counts the number of wavelengths over 2π, is termed the wavenumber. At any position the displacement u(x,t) is time-harmonic with time period T, where $T = 2\pi/\omega$. The circular frequency ω follows from Eq. (4.1) as

$$\omega = kc . \tag{4.2}$$

It follows that an alternative representation of u(x,t) is

$$u(x,t) = A \cos \left[\omega\left(\frac{x}{c} - t\right)\right] \tag{4.3}$$

Equations (4.1) and (4.3) represent trains of sinusoidal waves, which disturb at any instant of time the complete (unbouded) extent of the medium. Harmonic waves are steady-state waves, as opposed to transient waves (pulses).

If the phase velocity is independent of k, very short waves propagate with the same phase velocity as long waves. If the phase velocity does not depend on the wavelenght we say that the system is nondispersive. If the material is not purely elastic but displays dissipative behavior, it is found that the phase velocity of harmonic waves depends on the wavelength, and the system is said to be dispersive. Dispersion is an important phenomenon because it governs the change of shape of a pulse as it propagates through a dispersive medium. Dispersion occurs not only in inelastic bodies, but also in elastic waveguides, as will be discussed in Section 4.2.

For mathematical convenience we generally use, instead of Eq. (4.1), the expression

$$u(x,t) = A \exp[ik(x - ct)], \tag{4.4}$$

where $i = \sqrt{(-1)}$. Without stating it explicitly, henceforth it is understood that the real part of (4.4) is to be taken for the physical interpretation of the solution.

Standing waves

Let us consider two displacement waves of the same frequency and wavelength but traveling in opposite directions. Since the wave equation is linear the resultant displacement is

$$u(x,t) = A_+ e^{i(kx-\omega t+\gamma_+)} + A_- e^{i(kx+\omega t+\gamma_-)},$$

where A_+ and A_- are real-valued amplitudes, and γ_+ and γ_- are phase angles. If the amplitudes of the two simple harmonic waves are equal, $A_+ = A_- = A$, we can write

$$u(x,t) = 2A \exp[i(kx + \tfrac{1}{2}\gamma_+ + \tfrac{1}{2}\gamma_-)]\cos(\omega t - \tfrac{1}{2}\gamma_+ + \tfrac{1}{2}\gamma_-) .$$

The real part of this expression is

$$u(x,t) = 2A \cos(kx + \tfrac{1}{2}\gamma_+ + \tfrac{1}{2}\gamma_-) \cos(\omega t - \tfrac{1}{2}\gamma_+ + \tfrac{1}{2}\gamma_-) . \tag{4.5}$$

Equation (4.5) represents a standing wave, since the shape of the wave does not move. At points where $\cos(kx + \gamma_+/2 + \gamma_-/2) = 0$, the two traveling waves always cancel each other and the medium is at rest. These points are called the nodal points. Halfway between each pair of nodal points are the antinodes, where the motion has the largest amplitude.

Modes of free vibration

Standing waves of certain wavelenghts form the modes of free vibrations of an elastic body with external boundaries.

Propagation of pulses

As a consequence of the linearity of the wave propagation problems that are under discussion, it is allowable to express the total response to a number of separate excitations as the superposition of the individual responses. Linear superposition, in conjunction with integral representations of forcing functions provide us with the means of determining solutions to problems of transient elastic wave propagation.

If Eq. (4.4) satisfies the equation(s) governing wave propagation in the medium for the general case that the phase velocity c depends on the wavenumber (and thus on the frequency by virtue of $\omega = kc$), it is evident that

$$u(x,t) = \int_{-\infty}^{\infty} A^*(\omega)\, e^{i\omega[x/c(\omega)-t]}\, d\omega \tag{4.6}$$

also satisfies the governing equations. Equation (4.6) can, however, be interpreted as the solution to a boundary value problem for a semi-infinite medium $x \geq 0$ with boundary condition at $x = 0$.

$$u(0,t) = A(t) = \int_{-\infty}^{\infty} A^*(\omega)\, e^{-i\omega t} d\omega$$

If A(t) is known, its "exponential Fourier transform" $A^*(\omega)$ can be obtained from the well-known Fourier integral theorem

$$A(t) = \frac{1}{2\pi} \int_{-\infty}^{\infty} e^{-i\omega t} d\omega \int_{-\infty}^{\infty} A(s)\, e^{i\omega s}\, ds \ ,$$

i.e.,

$$A^*(\omega) = \frac{1}{2\pi} \int_{-\infty}^{\infty} A(s)\, e^{i\omega s}\, ds$$

4.2. Plane Harmonic Waves in a Homogeneous, Isotropic, Elastic Solid

It was shown in Part I that the displacement equation of motion for a homogeneous, isotropic, linearly elastic solid is of the form

$$\mu u_{i,jj} + (\lambda + \mu) u_{j,ji} = \rho \ddot{u}_i \tag{4.7}$$

As a simple extension of Eq. (4.1), a plane harmonic wave propagating in an arbitrary direction in an unbounded medium is represented by

$$u_i = A d_i \exp\,[\, ik(p_j x_j - ct)] \tag{4.8}$$

In Eq. (4.8), d_i and p_i are the components of unit vectors in the directions of motion and of propagation, respectively. The components of the position vector are denoted by x_i, and $p_i x_i$ = constant represents a plane normal to the unit vector with components p_i. Equation (4.8) thus represents a plane wave propagating in the direction of the propagation vector. By substituting Eq. (4.8) into the displacement equations of motion (4.7) we obtain after some manipulation

$$(\mu - \rho c^2)\, d_i + (\lambda + \mu)\,(p_j d_j)\, p_i = 0 . \tag{4.9}$$

Equation (4.9) is an equation for the phase velocity c. It is noted that the wave number k does not appear in (4.9), i.e., plane harmonic waves in an unbounded homogeneous isotropic linearly elastic medium are not dispersive.

Since p_i and d_i represent two different unit vectors, Eq. (4.9) can be satisfied in two ways only:

a) $p_i = \pm\, d_i$.

Consequently, $p_j d_j = \pm 1$, and Eq. (4.9) yields

$$c_L^2 = (\lambda + 2\mu)/\rho . \tag{4.10}$$

In this case the motion is parallel to the direction of propagation, and the wave is therefore called a longitudinal wave.

b) If $p_i \neq d_i$, both terms in (4.9) have to vanish independently, yielding

$$c_T^2 = \mu/\rho \quad \text{and} \quad p_j d_j = 0 \tag{4.11}$$

Now the motion is normal to the direction of propagation, and the wave is called a transverse wave.

Displacement potentials

Let us consider a representation of the displacement components in the form

$$u_i = \varphi_{,i} + e_{ijk}\, \psi_{k,j} \tag{4.12}$$

where e_{ijk} is the alternating tensor. Substitution of (4.12) into the displacement equation of motion yields that (4.7) is satisfied by (4.12) provided that

(4.13a,b) $$\varphi_{,ii} = \frac{1}{c_L^2}\ddot{\varphi} \quad \text{and} \quad \psi_{k,jj} = \frac{1}{c_T^2}\ddot{\psi}_k ,$$

where c_L and c_T are given by Eqs. (4.10) and (4.11), respectively. Equations (4.13a) and (4.13b) are uncoupled wave equations.

Although the scalar potential φ and the vector potential with components ψ_i are generally coupled through the boundary conditions, which still causes substantial mathematical complications, the use of the displacement decomposition generally simplifies the analysis. To determine the solution of a boundary-initial value problem one may simply select appropriate particular solutions of Eqs. (4.13a) and (4.13b) in terms of arbitrary functions or integrals over arbitrary functions. If these functions can subsequently be chosen so that the boundary conditions and the initial conditions are satisfied, then the solution to the problem has been found. The solution is unique by virtue of the uniqueness theorem of dynamic elasticity.

Waves in plane strain in an elastic layer

For motion in plane strain in the $(x_1 x_2)$-plane we have

$$u_3 \equiv 0, \quad \frac{\partial}{\partial x_3}(\;) \equiv 0.$$

Equations (4.12) then reduce to

(4.14) $$u_1 = \frac{\partial\varphi}{\partial x_1} + \frac{\partial\psi}{\partial x_2} ,$$

(4.15) $$u_2 = \frac{\partial\varphi}{\partial x_2} - \frac{\partial\psi}{\partial x_1} .$$

For simplicity of notation the subscript 3 has been omitted from ψ in (4.14) and (4.15). The relevant components of the stress tensor follows from Hooke's law as

(4.16) $$\tau_{21} = \mu\left(2\frac{\partial^2\varphi}{\partial x_1\partial x_2} - \frac{\partial^2\psi}{\partial x_1^2} + \frac{\partial^2\psi}{\partial x_2^2}\right)$$

$$\tau_{22} = \lambda\left(\frac{\partial^2\varphi}{\partial x_1^2} + \frac{\partial^2\varphi}{\partial x_2^2}\right) + 2\mu\left(\frac{\partial^2\varphi}{\partial x_2^2} - \frac{\partial^2\psi}{\partial x_1\partial x_2}\right) , \qquad (4.17)$$

while the potentials φ and ψ satisfy wave equations, which for plane strain are two-dimensional,

$$\frac{\partial^2\varphi}{\partial x_1^2} + \frac{\partial^2\varphi}{\partial x_2^2} = \frac{1}{c_L^2}\frac{\partial^2\varphi}{\partial t^2} , \qquad (4.18)$$

$$\frac{\partial^2\psi}{\partial x_1^2} + \frac{\partial^2\psi}{\partial x_2^2} = \frac{1}{c_T^2}\frac{\partial^2\psi}{\partial t^2} . \qquad (4.19)$$

To investigate wave motion in the elastic layer we consider solutions of (4.18) and (4.19) of the forms

$$\varphi = \Phi\,(x_2)\ \exp\,[\,i(kx_1 - \omega t)] , \qquad (4.20)$$

$$\psi = \Psi\,(x_2)\ \exp\,[\,i(kx_1 - \omega t)] . \qquad (4.21)$$

These expressions represent standing waves across the cross section, and progressive waves in the direction of the layer. Substituting (4.20) and (4.21) into (4.18) and (4.19), respectively, the solutions of the resulting equations are obtained as

$$\Phi\,(x_2) = A\sin\,(px_2) + B\cos\,(px_2) \qquad (4.22)$$

$$\Psi\,(x_2) = C\sin\,(qx_2) + D\cos\,(qx_2) , \qquad (4.23)$$

wherein

$$p^2 = \frac{\omega^2}{c_L^2} - k^2 , \quad q^2 = \frac{\omega^2}{c_T^2} - k^2 \qquad (4.24a,b)$$

By virtue of Eqs. (4.14) and (4.15), and Eqs. (4.22) and (4.23) the displacement components can be written in terms of elementary functions. For the displacement in the x_1-direction the motion is symmetric (antisymmetric) with regard to $x_2 = 0$, if u_1 contains cosines (sines). The displacement in the x_2-direction is symmetric (antisymmetric) if u_2 contains sines (cosines). The modes of wave propagation in the elastic layer may thus be split up into two systems of symmetric and antisymmetric modes, respectively.

The expressions for the displacements and the stresses must satisfy boundary conditions on the surfaces $x_2 = \pm h$. If the boundaries are free, we have at $x_2 = \pm h$, :

$$\tau_{21} = \tau_{22} = 0 .$$

These conditions yield the frequency equation, i.e., the equation relating the frequency ω to the wavenumber k. For the symmetric modes we obtain

$$\frac{\tan(qh)}{\tan(ph)} = -\frac{4k^2 pq}{(q^2 - k^2)^2} . \tag{4.25}$$

For the antisymmetric modes the boundary conditions yield

$$\frac{\tan(qh)}{\tan(ph)} = -\frac{(q^2 - k^2)^2}{4k^2 pq} . \tag{4.26}$$

Equations (4.25) and (4.26) are the well-known Rayleigh-Lamb frequency equations. These transcendental equations look deceptively simple. Although the frequency equations were derived at the end of the 19th century it was not until quite recently that the frequency spectrum was unraveled in complete detail by Mindlin, [1.12]. For a detailed discussion we refer to the book by Achenbach, [1.11]. For an arbitrarily specified value of the wavenumber k, Eqs. (4.25) and (4.26) yield an infinite number of solutions for the frequency ω. A specific wave motion of the layer, called a mode of propagation, corresponds to each solution satisfying (4.25) or (4.26). Thus, the frequency equation yields an infinite number of continuous curves, called branches, which display the relationship between the frequency ω and the wavenumber k for modes of wave propagation. The collection of branches constitutes the frequency spectrum.

4.3. Propagation of Harmonic Waves in Anisotropic Solids

Substitution of the effective stress-strain relations into the stress-equations of motion yields

$$C^{*}_{ijk\ell}\,\bar{u}_{k,\ell j} = \bar{\rho}\,\frac{\partial^2 \bar{u}_i}{\partial t^2} , \tag{4.27}$$

where $\bar{\rho}$ is an effective mass density. Let us consider

$$u_i = A\, d_i \exp[i\omega(x_j q_j - t)] \tag{4.28}$$

where ω is the (real-valued) angular frequency and the components of the slowness vector q_j are defined as

$$q_j = p_j / c,$$

Here p_j are the components of the vector which defines the direction of propagation and c is the phase velocity. Upon substitution of (4.28) into the displacement equation of motion we obtain

$$(C^*_{ijk\ell}\, q_j\, q_\ell - \rho\delta_{ik})\, d_k = 0 , \tag{4.29}$$

where δ_{ik} is the Kronecker delta. Equation (4.29) is an equation for the components d_k. For a nontrivial solution the determinant of the coefficients of d_k must vanish, which gives

$$\det|C^*_{ijk\ell}\, q_j\, q_\ell - \rho\delta_{ik}| = 0. \tag{4.30}$$

If we let

$$\Gamma_{ik} = C^*_{ijk\ell}\, p_j\, p_\ell ,$$

Eq. (4.30) can be rewritten as

$$\det|\Gamma_{ik} - \rho c^2 \delta_{ik}| = 0 . \tag{4.31}$$

The constants Γ_{ik} are known as the Christoffel stiffnesses. If the components p_j are given, Eq. (4.31) describes three velocity sheets in the space spanned up by p_j.

From the properties of the $C^*_{ijk\ell}$ it follows that Γ_{ik} is a symmetric and positive definite matrix. That is,

$$\Gamma_{ik} = \Gamma_{ki} , \quad \Gamma_{ik}\, d_i\, d_k \geqq 0 \quad \text{for all } d_i .$$

It follows that all of the eigenvalues of Γ_{ik} are real and positive and their corresponding eigenvectors are orthogonal. The physical interpretation of these observations is that for a given direction of wave propagation defined by p_i there will be three phase velocities c_I, c_{II} and c_{III}, and the three corresponding displacement vectors will be orthogonal. Contrary to the isotropic case the displacements are, however, neither truly longitudinal nor truly transverse in character.

An important observation is that the phase velocities are constants, so that harmonic waves propagating in an anisotropic medium are not dispersive. Thus, the effect of dispersion in an elastic composite cannot be described by the effective modulus theory.

Let us consider the case of transverse isotropy in somewhat more detail. A transversely isotropic material is defined by five elastic constants. Choosing the x_2-axis as the axis of symmetry the effective stress-strain relations are as stated in Section 3.3.

As shown earlier in this section it is generally rather complicated to study the propagation of plane time harmonic waves in an anisotropic medium. For a transversely isotropic material exceptions are waves propagating in the direction of the axis of symmetry, or in a direction normal to the axis of symmetry; these are very easy to analyze. For, example, if we consider a transverse wave motion defined by

$$u_2 = A \exp [\, ik(x_1 - ct)] \quad ,$$

we easily obtain

$$c = (C_{44}^{*}/\bar{\rho})^{\frac{1}{2}} \quad , \tag{4.31}$$

where $\bar{\rho}$ is defined as

$$\bar{\rho} = \frac{d_f}{d_f + d_m} \rho_f + \frac{d_m}{d_f + d_m} \rho_m \tag{4.32}$$

Similarly for a longitudinal wave motion defined by

$$u_1 = A \exp [\, ik(x_1 - ct)]$$

we find

$$c = (C_{11}^{*}/\bar{\rho})^{\frac{1}{2}} \tag{4.33}$$

The effective elastic constant C_{44}^{*} and C_{11}^{*} are defined by Eqs. (3.44) and (3.41), respectively.

4.4. Wave Motions in a Laminated Medium

We consider a laminated medium consisting of alternating plane, parallel layers of two homogeneous, isotropic materials. The Lamé elastic constants and the thicknesses of the high-modulus reinforcing layers and the low-modulus matrix layers are denoted by λ_f, μ_f, d_f and λ_m, μ_m, d_m, respectively, see Fig. 1.4.

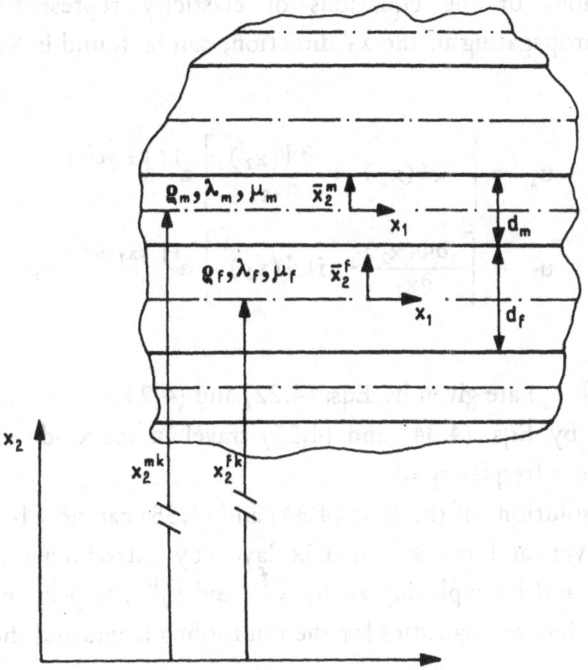

Fig. 1.4. Laminated composite.

Time-harmonic waves in an unbounded medium

For two-dimensional deformations associated with plane strain in the x_3-direction, wave motions representing waves propagating in the direction of the layering (x_1-direction) can be analyzed without much difficulty, using the equations of the theory of elasticity for all layers. For such waves the field quantities are of the form $F(x_2) \exp i(kx_1 - \omega t)$, where the function $F(x_2)$ has the same periodicity as the layering. As a consequence, the deformations are identical in all reinforcing layers and in all matrix layers, respectively. For this relatively simple state of deformation the attention can be restricted to one pair of neighboring layers. We choose the kth pair of reinforcing and matrix layers whose mid-plane positions are defined by x_2^{fk} and x_2^{mk}, respectively, see Fig. 1.4. For these two layers we define local coordinate systems $(x_1, \bar{x}_2^f, x_3)$ and $(x_1, \bar{x}_2^m, x_3)$, with axes parallel to x_i, and with origins in the mid-planes of the layers. In principle the problem at hand now consists of writing wave solutions of the equations of elasticity for the kth reinforcing layer and the kth matrix layer, and of ascertaining that the stresses and the displacements are continuous at the interfaces, and that the deformation pattern properly reflects the periodic structure of the medium.

Solutions of the equations of elasticity representing plane time-harmonic waves propagating in the x_1-direction, can be found in Section 4.2 in the forms

$$u_1 = \left[ik\Phi(x_2) + \frac{\partial\Psi(x_2)}{\partial x_2} \right] e^{i(kx_1-\omega t)} \tag{4.34}$$

$$u_2 = \left[\frac{\partial\Phi(x_2)}{\partial x_2} - ik\Psi(x_2) \right] e^{i(kx_1-\omega t)} , \tag{4.35}$$

where $\Phi(x_2)$ and $\Psi(x_2)$ are given by Eqs. (4.22) and (4.23)
Waves represented by Eqs. (4.34) and (4.35) travel in the x_1-direction with wave number k and circular frequency ω.

Wave solutions of the type (4.34) and (4.35) can now be written for the kth reinforcing layer and the kth matrix layer by introducing the appropriate material constants, and by replacing x_2 by $\bar{x}_2^f$ and $\bar{x}_2^m$, respectively. Subscripts f and m are used to denote quantities for the reinforcing layers and the matrix layers, respectively.

For deformations that are symmetric with respect to the mid-planes of the layers, we have $A_f = A_m = D_f = D_m = 0$. Referring to Fig. 3.1, the four conditions of continuity of the displacements and the stresses at the interface defined by $\bar{x}_2^f = \frac{1}{2} d_f$ and $\bar{x}_2^m = -\frac{1}{2} d_m$ yield four homogeneous equations for the four constants B_f, B_m, C_f, and C_m. Because it is possible to consider separately wave motions that are symmetric and antisymmetric with respect to the mid-planes of the individual layers, the displacements and the stresses at $\bar{x}_2^m = \frac{1}{2} d_m$ are matched with the corresponding displacements and stresses at $\bar{x}_2^f = -\frac{1}{2} d_m$, and the required periodicity of the deformation is achieved automatically.

The four homogeneous equations for the constants B_f, B_m, C_f and C_m yield non-trivial solutions only if the determinant of the coefficients vanishes. The latter requirement provides us with a relation between the frequency and the wavenumber, which is the frequency equation. It is a transcendental equation of the general form

$$F(\omega, k) = 0$$

For symmetric (longitudinal) waves propagating in the direction of the layering the frequency equation is stated explicitly in the paper by Achenbach and Herrmann,

[1.13].

For any value of the wavenumber k the frequency equation yields an infinite number of solutions for the frequency ω. A specific wave motion, called a mode of propagation, corresponds to each of these frequencies. The curves which display the relationships between the frequency and the wavenumber for the various modes are the branches of the frequency spectrum. The frequencies of the three lowest symmetric modes are shown in Fig. 1.5 for real-valued wavenumbers. It is to be expected that the frequency equation also yields solutions for wavenumbers that are complex. The corresponding modes, which are of interest in the analysis of transient wave motions and in the discussion of motions near boundaries, have not yet been analyzed. In Fig. 1.5 we have plotted the dimensionless frequency Ω, and the dimensionless wavenumber ξ. These quantities are defined as

$$\Omega = \frac{\omega d_f}{(\mu_m/\rho_m)^{\frac{1}{2}}} \ , \quad \text{and} \ \xi = kd_f$$

respectively. The values of the material and the geometrical parameters were chosen as

$$\frac{\mu_f}{\mu_m} = 50 \ , \quad \frac{\rho_f}{\rho_m} = 3 \ , \quad \frac{d_f}{d_f + d_m} = 0.8 \ ,$$

and

$$\nu_f = 0.3 \ , \quad \nu_m = 0.35 \ ,$$

where μ, ρ and ν denote a shear modulus, a mass density and a Poisson's ratio, respectively.

It is of interest to consider some limit cases of the frequency equation. For $\mu_f = \mu_m$ and $\rho_f = \rho_m$, i.e., for a homogeneous isotropic medium, the frequency equation reduces to a simple equation which yields as solutions the well-known frequencies for longitudinal and transverse waves in an unbounded isotropic medium. In the limit of vanishing ξ, i.e., for infinitely long waves, the frequency equation splits up into the product of two equations governing the frequencies of uncoupled symmetric thickness shear and symmetric thickness stretch motions, respectively. In certain ranges of the material constants the cut-off frequency of the lowest symmetric thickness shear mode is less than the cut-off frequency of the lowest symmetric thickness stretch mode; this observation has to be taken into account when constructing approximate theories. For more details on

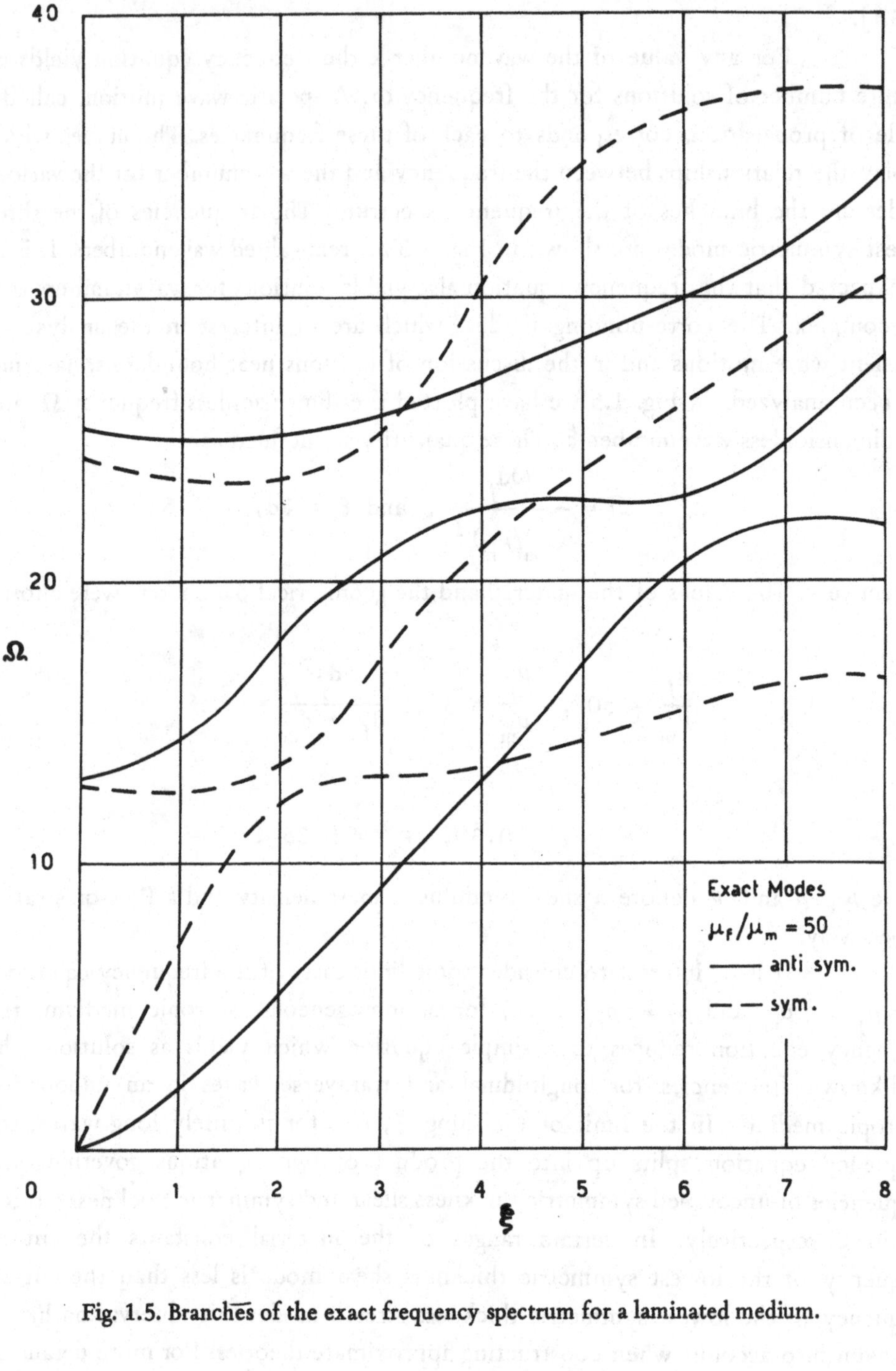

Fig. 1.5. Branches of the exact frequency spectrum for a laminated medium.

these matters we refer to Achenbach and Herrmann, [1.13]. At vanishing wavenumber, the frequency of the lowest symmetric mode vanishes, and we can compute the slope of the $\Omega-\xi$ curve at $\xi = 0$ from the frequency equation. This slope is equal to the limiting phase velocity of the lowest mode, and we find

$$c = (C_{11}^{*} / \bar{\rho})^{\frac{1}{2}} , \tag{4.7}$$

where C_{11}^{*} and $\bar{\rho}$ are defined by Eqs. (3.41) and (4.32). Thus, the phase velocity of the lowest mode at $\xi = 0$ according to the exact theory is just equal to the phase velocity of the single mode of longitudinal motion according to the effective modulus theory.

The analysis runs along very similar lines for wave motions with displacements that are antisymmetric with respect to the mid-planes of the layers. The complete frequency equation for this case is presented in a paper by Achenbach, [1.14]. The frequencies for the three lowest antisymmetric modes are also plotted in Fig. 1.5. In investigating the frequency eqation for antisymmetric motion in the limit of vanishing ξ we find equations governing the frequencies of simple antisymmetric thickness stretch and thickness shear motions. The phase velocity at vanishing wavenumber of the lowest antisymmetric mode can be obtained as

$$c = (C_{44}^{*} / \bar{\rho})^{\frac{1}{2}} , \tag{4.36}$$

where C_{44}^{*} and $\bar{\rho}$ are defined by Eqs. (3.44) and (4.32) respectively. Equation (4.36) is the phase velocity according to the effective modulus theory for transverse waves propagating in the direction of the layering, see Eq. (4.31). It is illustrative to plot the phase velocity of the lowest antisymmetric mode for various values of μ_f/μ_m versus the dimensionless wavenumber ξ. These phase velocities are shown in Fig. 1.6. As ξ increases the phase velocities show a considerable deviation from the constant values according to the effective modulus theory, especially for higher values of μ_f/μ_m.

Plane harmonic waves propagating in an arbitrary direction were examined by Sve, [1.15]. Some of Sve's results are plotted in Fig. 1.7. The frequency spectrum shown in Fig. 1.7 clearly reveals the different nature of the propagation of sinusoidal waves at various angles of incidence. For waves propagating normal to the layering there are frequency bands in which no waves with real-valued wavenumbers can propagate. This indicates that the laminated medium acts as a wavefilter for waves propagating normal to the layering. For propagation not normal to the layering a wave with any frequency can propagate

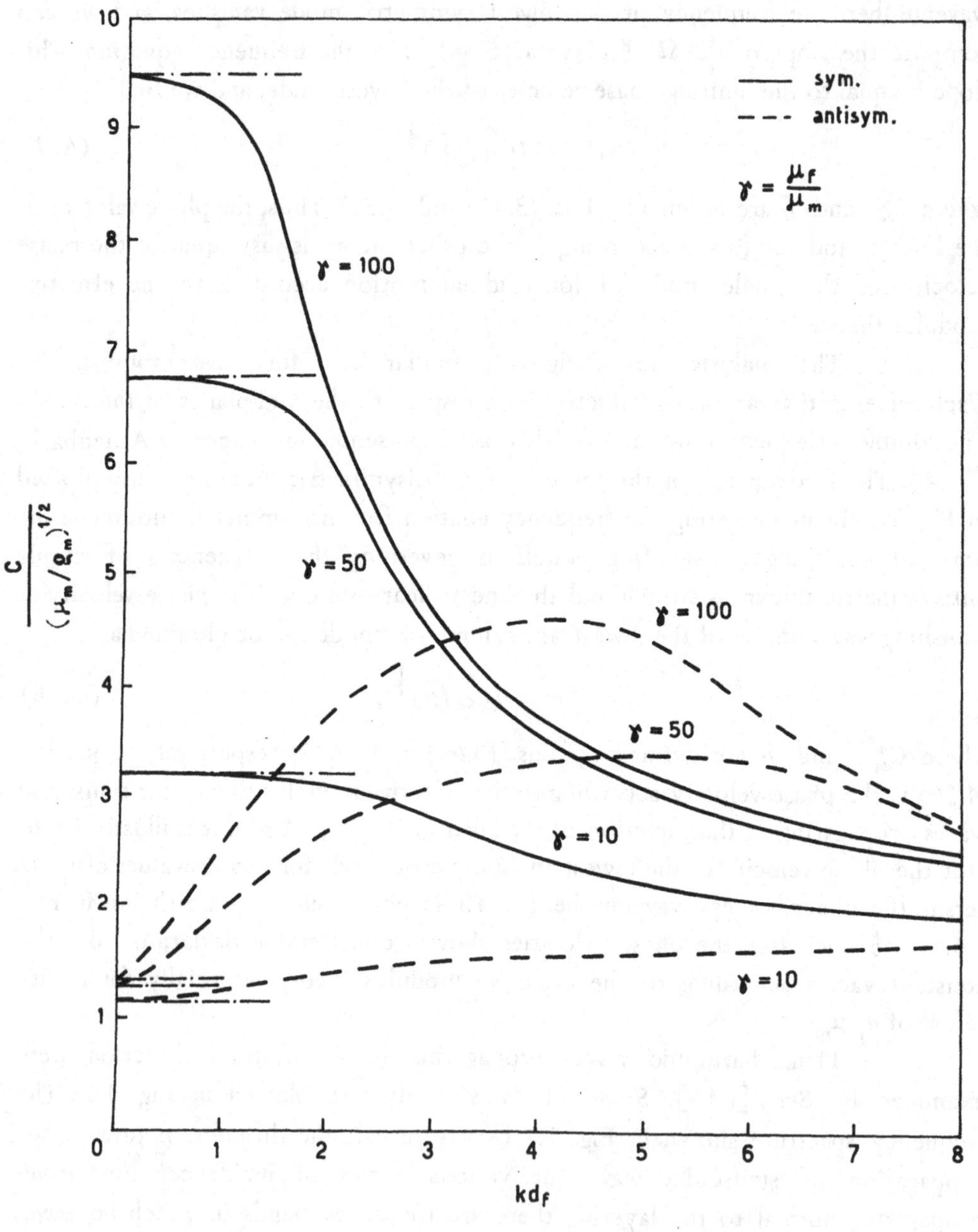

Fig. 1.6. Phase velocity versus the wavenumber for the lowest symmetric and antisymmetric modes in the laminated medium.

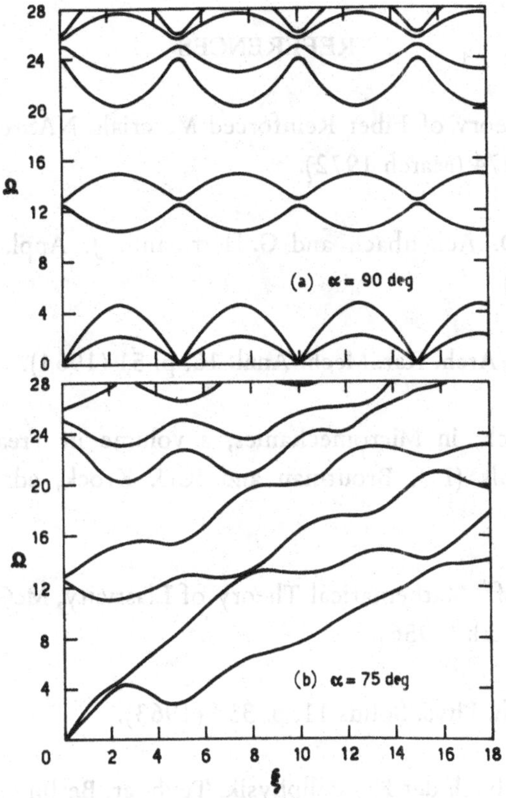

Fig. 1.7. Branches of the exact frequency spectrum for oblique and normal incidence; $\mu_f/\mu_m = 50$ (from Sve, 1971).

with a real-valued wavenumber.

In summary, the exact analysis of the propagation of time-harmonic waves in a laminated medium of the type shown in Fig. 1.5 reveals an infinite number of symmetric and antisymmetric modes. The effective modulus theory yields one symmetric (longitudinal) and one antisymmetric (transverse) mode, whose constant phase velocities are equal to the limiting phase velocities at vanishing wavenumber (infinite wavelength) of the lowest symmetric and antisymmetric modes, respectively, of the exact spectrum.

REFERENCES

[1.1] Z. Hashin, Theory of Fiber Reinforced Materials, NASA Contractor Report No. 1974 (March 1972).

[1.2] C.T. Sun, J.D. Achenbach and G. Herrmann, J. Appl. Mech. 35, p. 408 (1968).

[1.3] R. D. Mindlin, Arch. Rat. Mech. Anal. 16, p. 51 (1964).

[1.4] J.D. Achenbach, in Micromechanics, a volume in Treatise of Composite Materials (L.J. Broutman and R.H. Krock, eds.) Academic Press, (1974).

[1.5] I.S. Sokolnikoff, Mathematical Theory of Elasticity, McGraw Hill Book Co., New York (1956).

[1.6] R. Hill, J. Mech. Phys. Solids 11, p. 357 (1963).

[1.7] W. Voigt, Lehrbuch der Krystallphysik, Teubner, Berlin (1910).

[1.8] A. Reuss, Z. Angew. Math. Mech. 9, p. 49 (1929).

[1.9] G. W. Postma, Geophysics XX, p. 780 (1955).

[1.10] Y. C. Fung, Foundations of Solid Mechanics, Prentice-Hall Inc. (1965).

[1.11] J. D. Achenbach, Wave Propagation in Elastic Solids, North-Holland American Elsevier, Amsterdam/New York (1973).

[1.12] R. D. Mindlin, in Structural Mechanics (J. N. Goodier and N. J. Hoff, eds.), p. 199, Pergamon Press, London (1960).

[1.13] J.D. Achenbach and G. Herrmann, in Dynamics of Structured Solids (G. Herrmann, ed.), p. 23, American Society of Mechanical Engineers, New York (1968).

[1.14] J. D. Achenbach, J. Acoust. Soc. Am. 43, p. 1451 (1968).

[1.15] C. Sve, J. Appl. Mech. 38, p. 477 (1971).

PART II

THE DIRECTIONALLY REINFORCED COMPOSITE AS A HOMOGENEOUS CONTINUUM WITH MICROSTRUCTURE

CHAPTER 1

THE LAMINATED MEDIUM

1.1. Introduction

For most solids used in technological applications, the lengths characterizing the ever-present inhomogeneities are much smaller than lengths characterizing the deformation. For the purpose of static and dynamic stress analysis it is then justifiable to describe the mechanical behavior by a classical, homogeneous, isotropic or anisotropic continuum. The corresponding theory, which is termed the effective modulus theory, was discussed in some detail in Part I.

For materials that are fabricated by compounding reinforcing elements and a matrix material to form a directionally reinforced composite material, the characteristic lenghts of the medium are substantially larger. Although for most loading conditions the effective modulus theory will still be satisfactory for such composites, it is conceivable that under certain loading conditions, particularly those generating a dynamic response for which the characteristic lengths of the deformation are small, a classical continuum will not give an altogether satisfactory description of the mechanical behavior. For example, the effect of dispersion of time-harmonic waves in an unbounded medium is not described by the effective modulus theory. Dispersion may, however, be of significance in a directionally reinforced composite since it governs the change of shape of a propagating pulse.

For a laminated medium the above considerations have motivated the formulation of a homogeneous continuum model which can describe dynamic effects due to the structuring of the inhomogeneous laminated medium. This Part is devoted to a detailed discussion of a theory for such a homogeneous continuum model. The particular model that is considered here was developed in Refs. 2.1 and 2.2 .

The system of governing equations is derived in two stages. The first stage of the derivation involves certain assumptions and operations within the discrete system of layers. In particular, it is assumed that the motions of the individual layers can be described by two-term expansions in a system of local coordinates. The kinematic variables that are introduced in the expansions are defined at the midplanes of the layers only. On the basis of these assumptions, strain and kinetic energies for representative elements of the layers are computed. In the next stage of the derivation a transition is made from the system of discrete layers to the homogeneous continuum model. The transition is accomplished by defining fields for the kinematical and dynamical variables that are continuous in the coordinate normal to the layering. In discrete planes, which are the midplanes of the layers, the continuous field variables assume the same values as the corresponding field variables that were defined in the discrete system of layers. After averaging, the previously computed kinetic and strain energies can then be interpreted as kinetic and strain energy densities. A subsequent application of Hamilton's principle, where certain continuity relations are included through the use of Lagrangian multipliers, subsequently yields a set of displacement equations of motion. The results of 2.1 demonstrated the advantages of the theory for dynamic problems and justified a study of further ingredients of a complete theory, such as constitutive equations, boundary conditions, uniqueness, etc. These aspects of the theory were discussed in [2.2].

The system of linear field equations, which consists of balance equations, constitutive relations and a constraint condition, resembles the equations of a linear theory of elasticity with microstructure. The resemblance is briefly explored.

1.2. Kinematics and Displacement Equations of Motion

We consider a laminated medium consisting of alternating layers of two homogeneous materials, see Fig. 2.1. It is specified that the field variables and the material parameters in the material whose resistance to deformation is higher (the high-modulus or reinforcing layers) are denoted by subscripts and superscripts f (fiber). The corresponding quantities in the other layers (the low-modulus or matrix layers) are denoted by subscripts and superscripts m (matrix). We focus attention on an element of the k-th pair of reinforcing and matrix layers whose midplane positions are defined by $x_2^{(fk)}$ and $x_2^{(mk)}$, respectively, see Fig. 2.2, and we define two local Cartesian coordinate systems

$$\left(x_1, \bar{x}_2^{(f)}, x_3\right) \text{ and } \left(x_1, \bar{x}_2^{(m)}, x_3\right).$$

If the characteristic length of the deformation is larger than the thickness of the layers, then, analogously to plate theories, the displacements in the kth pair of layers can be approximated by

$$u_i^{(fk)} = \bar{u}_i^{(fk)}\left(x_1, x_2^{(fk)}, x_3, t\right) + \bar{x}_2^{(f)} \psi_{2i}^{(fk)}\left(x_1, x_2^{(fk)}, x_3, t\right) \tag{1.1}$$

$$u_i^{(mk)} = \bar{u}_i^{(mk)}\left(x_1, x_2^{(mk)}, x_3, t\right) + \bar{x}_2^{(m)} \psi_{2i}^{(mk)}\left(x_1, x_2^{(mk)}, x_3, t\right) , \tag{1.2}$$

where i = 1,2, or 3, and where $\bar{u}_i^{(fk)}$ and $\bar{u}_i^{(mk)}$ are the displacements in the midplane of the kth reinforcing and kth matrix layer, respectively. In Eq (1.1), $\psi_{21}^{(fk)}$ and $\psi_{23}^{(fk)}$ represent antisymmetric thickness shear deformations, and $\psi_{22}^{(fk)}$ represents symmetric thickness stretch deformation of the kth reinforcing layer. Analogous definitions apply to $\psi_{21}^{(mk)}$, $\psi_{23}^{(mk)}$ and $\psi_{22}^{(mk)}$. Thus the displacements in a layers are expressed as the displacements in the midplane plus additional terms which increase linearly with the distance from the midplane. Note that $\bar{u}_i^{(fk)}$, $\bar{u}_i^{(mk)}$, $\psi_{2i}^{(mk)}$ are defined at discrete values of x_2, but they are continuous functions of x_1, x_3 and t.

Since the displacements are continuous across the interface of the kth pair of layers, we can write

$$\bar{u}_i^{(mk)}\left(x_1, x_2^{(mk)}, x_3, t\right) - \bar{u}_i^{(fk)}\left(x_1, x_2^{(fk)}, x_3, t\right) = \frac{1}{2} d_f \psi_{2i}^{(fk)}\left(x_1, x_2^{(fk)}, x_3, t\right) + $$

$$+ \frac{1}{2} d_m \psi_{2i}^{(mk)}\left(x_1, x_2^{(mk)}, x_3, t\right) \tag{1.3}$$

where, as shown in Fig. 2.2, d_f and d_m are the thicknesses of the reinforcing and matrix layers, respectively. The continuity of the stresses at the interfaces comes up at a later stage, when we discuss the balance of momentum equations.

With a view toward constructing representative potential and kinetic energy densities for the laminated medium, we first compute expressions for the strain energy based on the displacement approximations (1.1)-(1.3). In an isotropic linearly elastic body the strain energy density can be written as

$$W = \frac{1}{2} (\lambda + 2\mu)\left(\epsilon_{11}^2 + \epsilon_{22}^2 + \epsilon_{33}^2\right) + \lambda\left(\epsilon_{11}\epsilon_{22} + \epsilon_{11}\epsilon_{33} + \epsilon_{22}\epsilon_{33}\right) +$$

$$+ 2\mu\left(\epsilon_{12}^2 + \epsilon_{23}^2 + \epsilon_{13}^2\right) , \tag{1.4}$$

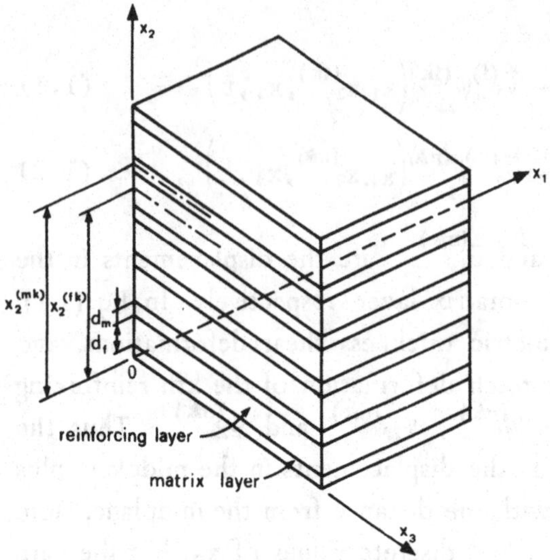

Fig. 2.1. Layered Medium

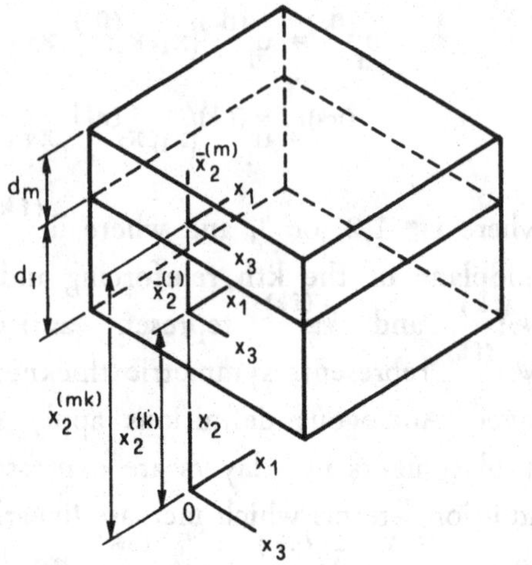

Fig. 2.2. Element of k-th pair of layers

where λ and μ are Lamé's elastic constants and the components of the strain tensor ϵ_{ij} are defined by

$$\epsilon_{ij} = \frac{1}{2}(\partial_j u_i + \partial_i u_j), \tag{1.5}$$

in which

$$\partial_j u_i = \partial u_i / \partial x_j \tag{1.6}$$

Substituting Eqs. (1.1) into (1.5), where, however, the differentiation in the x_2-direction should be with respect to the local coordinate $\bar{x}_2^{(f)}$, we find

$$\epsilon_{\alpha\beta}^{(fk)} = \frac{1}{2}\left(\partial_\beta \bar{u}_\alpha^{(fk)} + \partial_\alpha \bar{u}_\beta^{(fk)}\right) + \frac{1}{2}\bar{x}_2^{(f)}\left(\partial_\beta \psi_{2\alpha}^{(fk)} + \partial_\alpha \psi_{2\beta}^{(fk)}\right) \tag{1.7}$$

$$\epsilon_{2\beta}^{(fk)} = \epsilon_{\beta 2}^{(fk)} = \frac{1}{2}\left(\partial_\beta \bar{u}_2^{(fk)} + \psi_{2\beta}^{(fk)}\right) + \frac{1}{2}\bar{x}_2^{(f)}\,\partial_\beta \psi_{22}^{(fk)} . \tag{1.8}$$

$$\epsilon_{22}^{(fk)} = \psi_{22}^{(fk)} \tag{1.9}$$

In Eqs. (1.7)-(1.9), the Greek indices α and β can assume the values 1 or 3 only. By replacing superscripts f by superscripts m in Eqs. (1.7)-(1.9) the corresponding expressions for the strain components in the kth matrix layer are obtained.

Substitution of the expressions (1.7)-(1.9) into Eq. (1.4) and integration of the resulting expression over the thickness d_f yields the strain energy stored per unit surface area in the $(x_1 x_3)$-plane of the kth reinforcing layer as

$$
\begin{aligned}
W^{(fk)} = \frac{1}{2}\, d_f \left(\lambda_f + 2\mu_f\right) & \left[\left(\partial_1 \bar{u}_1^{(fk)}\right)^2 + \left(\partial_3 \bar{u}_3^{(fk)}\right)^2 + \left(\psi_{22}^{(fk)}\right)^2\right] \\
& + d_f \lambda_f \left(\partial_1 \bar{u}_1^{(fk)} \partial_3 \bar{u}_3^{(fk)} + \partial_1 \bar{u}_1^{(fk)} \psi_{22}^{(fk)} + \partial_3 \bar{u}_3^{(fk)} \psi_{22}^{(fk)}\right) \\
& + \frac{1}{2}\, d_f \mu_f \left[\left(\partial \bar{u}_2^{(fk)} + \psi_{21}^{(fk)}\right)^2 + \left(\partial_3 u_2^{(fk)} + \psi_{23}^{(fk)}\right)^2\right. \\
& \left. + \left(\partial_3 \bar{u}_1^{(fk)} + \partial_1 \bar{u}_3^{(fk)}\right)^2\right] \\
& + \frac{1}{24}\, d_f^3 \left(\lambda_f + 2\mu_f\right) \left[\left(\partial_1 \psi_{21}^{(fk)}\right)^2 + \left(\partial_3 \psi_{23}^{(fk)}\right)^2\right] \\
& + \frac{1}{12}\, d_f^3 \lambda_f \partial_1 \psi_{21}^{(fk)} \partial_3 \psi_{23}^{(fk)} \\
& + \frac{1}{24}\, d_f^3 \mu_f \left[\left(\partial_1 \psi_{22}^{(fk)}\right)^2 + \left(\partial_3 \psi_{22}^{(fk)}\right)^2 + \left(\partial_3 \psi_{21}^{(fk)} + \partial_1 \psi_{23}^{(fk)}\right)^2\right]
\end{aligned}
\tag{1.10}
$$

A similar computation yields an expression for the strain energy stored in an element of unit surface area of the kth matrix layer. The actual expression for $W^{(mk)}$ can be written by replacing in Eq. (1.10) subscripts and superscripts f by subscripts and superscripts m, respectively. The total strain energy stored in a cell of thickness $d_f + d_m$ and of unit surface area in the $(x_1 x_3)$-plane then is

$$
W^{(k)} = W^{(fk)} + W^{(mk)} \tag{1.11}
$$

The average over the volume of the cell is

$$(1.12)\qquad W^{(k)}_{ave} = \frac{W^{(fk)} + W^{(mk)}}{d_f + d_m}$$

The kinetic energy density of a continuum is defined by

$$(1.13)\qquad T = \frac{1}{2}\rho\left[(\dot{u}_1)^2 + (\dot{u}_2)^2 + (\dot{u}_3)^2\right],$$

where ρ is the mass density. By the use of the displacement expansion (1.1) the kinetic energy per unit surface area of the reinforcing material in cell k is obtained as

$$(1.14)\qquad T^{(fk)} = \sum_{i=1}^{3} \frac{1}{2}\rho_f d_f\left\{\left(\dot{\bar{u}}_i^{(fk)}\right)^2 + \frac{1}{12}\left(d_f \dot{\psi}_{2i}^{(fk)}\right)^2\right\}$$

The kinetic energy stored in the matrix material of the cell k is obtained in a similar manner. The total kinetic energy stored in cell k is

$$(1.15)\qquad T^{(k)} = T^{(fk)} + T^{(mk)}$$

The average over the volume of cell k is

$$(1.16)\qquad T^{(k)}_{ave} = \frac{T^{(fk)} + T^{(mk)}}{d_f + d_m}$$

The motion of the reinforcing layers and the matrix layers is now described by the field variables $\bar{u}_i^{(fk)}$, $\psi_{2i}^{(fk)}$, and $\bar{u}_i^{(mk)}$, $\psi_{2i}^{(mk)}$, respectively. These field variables are defined only on discrete parallel planes $x_2 = x_2^{(fk)}$ and $x_2 = x_2^{(mk)}$, respectively. As a first step toward the construction of a homogeneous continuum model for the laminated medium we now introduce fields that are continuous functions of x_2 and whose values at $x_2 = x_2^{(fk)}$ and $x_2 = x_2^{(mk)}$ coincide with the actual field variables in the midplanes of the layers. The step is indicated symbolically by writing $\bar{u}_i^{(m)}(x_i, t)$ rather than $\bar{u}_i^{(mk)}(x_1, x_2^{(mk)}, x_3, t)$, etc. In this manner, four continuous fields are introduced:

$$\bar{u}_i^{(f)}(x_i, t),\ \bar{u}_i^{(m)}(x_i, t),\ \psi_{2i}^{(f)}(x_i, t) \text{ and } \psi_{2i}^{(m)}(x_i, t).$$

The number of four field variables is, however, reduced to three by the argument that $\bar{u}_i^{(f)}(x_i, t)$ and $\bar{u}_i^{(m)}(x_i, t)$ should be considered as representing the same function (the "gross displacement") in different regions, namely, in the reinforcing layers and the matrix layers, respectively. We thus replace $\bar{u}_i^{(f)}(x_i, t)$ and $\bar{u}_i^{(m)}(x_i, t)$ by the one gross displacement $\bar{u}_i(x_i, t)$. The functions $\psi_{2i}^{(f)}(x_i, t)$ and $\psi_{2i}^{(m)}(x_i, t)$ describe "local deformations" which remain distinguished in he reinforcing and the matrix layers.

The local deformations are, however, related to the gross displacement through the conditions (1.3), representing continuity of the displacement at the interfaces. Nothing that $x_2^{(mk)} = x_2^{(fk)} + (d_m + d_f)/2$, and assuming that the thicknesses of the layers are sufficiently small, the difference relations (1.3) can be replaced by the differential relations

$$\partial_2 \bar{u}_i(x_i, t) = \eta \psi_{2i}^{(f)}(x_i, t) + (1-\eta)\psi_{2i}^{(m)}(x_i, t), \tag{1.17}$$

where $\partial_j \bar{u}_i(x_i, t) = \partial \bar{u}_i(x_i, t)/\partial x_j$, and η is defined as

$$\eta = d_f/(d_f + d_m). \tag{1.18}$$

The interface conditions in the system of discrete layers, as given by (1.3), have thus been turned into a constraint condition which holds at any point in the continuum. The limiting process involved in the passage from (1.3) to (1.18) is formally justifiable in the limit of $d_f \to 0$ and $d_m \to 0$, but keeping η constant. Here we assume without rigorous mathematical justification that the passage is also valid for finite values of d_f and d_m. This procedure is followed in an attempt to construct a continuum model for the layered medium which shows the same gross dynamic behavior and whose virtues will be verified in the sequel.

As a direct implication of the introduction of a continuous field for the gross displacement and two continuous fields of local deformations, we can also introduce a strain energy density W and a kinetic energy density T. These densities are continuous functions of x_i and t, and their forms follow from Eqs (1.12) and (1.14), respectively. For example, for the kinetic energy density we find

$$T = \sum_{i=1}^{3} \left\{ \frac{1}{2}\rho\left(\dot{\bar{u}}_i\right)^2 + \frac{1}{2}\eta I_f\left(\dot{\psi}_{2i}^{(f)}\right)^2 + \frac{1}{2}(1-\eta) I_m\left(\dot{\psi}_{2i}^{(m)}\right)^2 \right\}, \tag{1.19}$$

where

$$\rho = \eta\rho_f + (1-\eta)\rho_m \tag{1.20}$$

$$I_f = \frac{1}{12} d_f^2 \rho_f \tag{1.21}$$

and

$$I_m = \frac{1}{12} d_m^2 \rho_m \tag{1.22}$$

The strain energy density can be written analogously.

At this stage we have constructed a strain energy density as an expression in terms of local deformations and the gradients of the local deformations and the gross displacements. A kinetic energy density has been obtained in terms of the first order time derivatives of the gross displacements and the local deformations. Considering a fixed regular region V of the medium, the displacement equations of motion can then be obtained by invoking Hamilton's principle for independent variations of the dependent field quantities in V and in a specified time interval $t_o \leq t \leq t_1$. A statement of Hamilton's principle can be found in Ref. 1.11 . Here we are interested only in the displacement equations of motion and we restrict the admissible variations to ones that vanish identically on the bounding surface of V. In the absence of body forces the variational problem then reduces to finding the Euler equations for

$$\delta \int_{t_o}^{t} \int_V F \, dt dV = 0 , \tag{1.23}$$

where the functional F is defined as

$$F = T - W$$

An elegant and convenient method of taking the continuity conditions (1.17) into account is to introduce them as subsidiary conditions through the use of Lagrangian multipliers. The variational problem may then be redefined by using the functional

$$F = T - W - \Sigma_{2i} S_i \tag{1.24}$$

in Eq. (1.23), where the Lagrangian multipliers Σ_{2i} are functions of x_j and t, and

$$S_i = (d_f + d_m)\partial_2 \bar{u}_i - d_f \psi_{2i}^{(f)} - d_m \psi_{2i}^{(m)} \tag{1.25}$$

Since F depends only on the dependent field variables and their first order derivatives the system of Euler equations may be written as

$$\sum_{r=1}^{4} \frac{\partial}{\partial q_r} \left[\frac{\partial F}{\partial(\partial f_s / \partial q_r)} \right] - \frac{\partial F}{\partial f_s} = 0 \tag{1.26}$$

In Eq. (1.26) f_s represents the twelve dependent variables $\bar{u}_i$, $\psi_{2i}^{(f)}$, $\psi_{2i}^{(m)}$ and Σ_{2i}, and q_r are the spatial variables x_i and time t. The twelve displacement equations of motion which follow from (1.26) and (1.24) are

$$a_1 \partial_{11} \bar{u}_1 + a_2 \partial_{33} \bar{u}_1 + (a_2 + a_3)\partial_{13} \bar{u}_3 + a_4 \partial_1 \psi_{22}^{(f)} + a_{10} \partial_1 \psi_{22}^{(m)} + \partial_2 \Sigma_{21} = \rho \ddot{\bar{u}}_1 \qquad (1.27)$$

$$a_2 \partial_{11} \bar{u}_2 + a_2 \partial_{33} \bar{u}_2 + a_8 \partial_1 \psi_{21}^{(f)} + a_8 \partial_3 \psi_{23}^{(f)} + a_{14} \partial_1 \psi_{21}^{(m)} + a_{14} \partial_3 \psi_{23}^{(m)} + \partial_2 \Sigma_{22} = \rho \ddot{\bar{u}}_2 \qquad (1.28)$$

$$a_1 \partial_{33} \bar{u}_3 + a_2 \partial_{11} \bar{u}_3 + (a_2 + a_3)\partial_{13} \bar{u}_1 + a_4 \partial_3 \psi_{22}^{(f)} + a_{10} \partial_3 \psi_{22}^{(m)} + \partial_2 \Sigma_{23} = \rho \ddot{\bar{u}}_3 \qquad (1.29)$$

$$a_6 \partial_{11} \psi_{21}^{(f)} + (a_7 + a_9)\partial_{13} \psi_{23}^{(f)} + a_9 \partial_{33} \psi_{21}^{(f)} - a_8 \partial_1 \bar{u}_2 - a_8 \psi_{21}^{(f)} + \eta \Sigma_{21} = \eta \, I_f \ddot{\psi}_{21}^{(f)} \qquad (1.30)$$

$$a_9 \partial_{11} \psi_{22}^{(f)} + a_9 \partial_{33} \psi_{22}^{(f)} - a_4 \partial_1 \bar{u}_1 - a_4 \partial_3 \bar{u}_3 - a_5 \psi_{22}^{(f)} + \eta \Sigma_{22} = \eta \, I_f \ddot{\psi}_{22}^{(f)} \qquad (1.31)$$

$$a_6 \partial_{33} \psi_{23}^{(f)} + (a_7 + a_9)\partial_{13} \psi_{21}^{(f)} + a_9 \partial_{11} \psi_{23}^{(f)} - a_8 \partial_3 \bar{u}_2 - a_8 \psi_{23}^{(f)} + \eta \Sigma_{23} = \eta \, I_f \ddot{\psi}_{23}^{(f)} \qquad (1.32)$$

$$a_{12} \partial_{11} \psi_{21}^{(m)} + (a_{13} + a_{15})\partial_{13} \psi_{23}^{(m)} + a_{15} \partial_{33} \psi_{21}^{(m)} - a_{14} \partial_1 \bar{u}_2 - a_{14} \psi_{21}^{(m)} + (1-\eta)\Sigma_{21} = (1-\eta) I_m \ddot{\psi}_{21}^{(m)} \qquad (1.33)$$

$$a_{15}\,\partial_{11}\,\psi_{22}^{(m)} + a_{15}\,\partial_{33}\,\psi_{22}^{(m)} - a_{10}\,\partial_1\bar{u}_1 - a_{10}\,\partial_3\bar{u}_3 - a_{11}\,\psi_{22}^{(m)} + (1-\eta)\Sigma_{22} = (1-\eta)\,I_m\,\ddot{\psi}_{22}^{(m)} \tag{1.34}$$

$$a_{12}\,\partial_{33}\,\psi_{23}^{(m)} + (a_{13} + a_{15})\,\partial_{13}\,\psi_{21}^{(m)} + a_{15}\,\partial_{11}\,\psi_{23}^{(m)} - a_{14}\,\partial_3\bar{u}_2 - a_{14}\,\psi_{23}^{(m)} + (1-\eta)\Sigma_{23} = (1-\eta)\,I_m\,\ddot{\psi}_{23}^{(m)} \tag{1.35}$$

$$\partial_2\bar{u}_1 - \eta\psi_{21}^{(f)} - (1-\eta)\psi_{21}^{(m)} = 0 \tag{1.36}$$

$$\partial_2\bar{u}_2 - \eta\psi_{22}^{(f)} - (1-\eta)\psi_{22}^{(m)} = 0 \tag{1.37}$$

$$\partial_2\bar{u}_3 - \eta\psi_{23}^{(f)} - (1-\eta)\psi_{23}^{(m)} = 0, \tag{1.38}$$

where the constants $a_1 \ldots a_{15}$ are defined as

$$a_1 = \eta(\lambda_f + 2\mu_f) + (1-\eta)(\lambda_m + 2\mu_m)\,;\quad a_2 = \eta\mu_f + (1-\eta)\mu_m\,;$$

$$a_3 = \eta\lambda_f + (1-\eta)\lambda_m\,;\quad a_4 = \eta\lambda_f\,;\quad a_5 = \eta(\lambda_f + 2\mu_f)\,;$$

$$a_6 = d_f^2\,\eta(\lambda_f + 2\mu_f)/12\,;\quad a_7 = d_f^2\,\eta\lambda_f/12\,;\quad a_8 = \eta\mu_f\,;$$

$$a_9 = d_f^2\,\eta\mu_f/12\,;\quad a_{10} = (1-\eta)\lambda_m\,;\quad a_{11} = (1-\eta)(\lambda_m + 2\mu_m)\,;$$

$$a_{12} = d_m^2(1-\eta)(\lambda_m + 2\mu_m)/12\,;\quad a_{13} = d_m^2(1-\eta)\lambda_m/12\,;$$

$$a_{14} = (1-\eta)\mu_m\,;\quad a_{15} = d_m^2(1-\eta)\mu_m/12\,.$$

1.3. Correspondence with Theories of Elasticity with Microstructure

The development of generalized theories for elastic continua sprang from the desire to describe various phenomena on the microscale which cannot be accounted for by classical continuum mechanics. The first formal theory of this type was probably constructed by E. and F. Cosserat,[2.3], whose theory entailed the introduction of the couple per unit area, acting across a surface within a material volume or on its boundary, in addition to the usual force per unit area. A modern derivation of a Cosserat-type theory and a discussion of typical effects of couple stresses were given by Mindlin and Tiersten,[2.4]. The derivation of the theory of

Ref. 2.4 was within the framework of a linearized form of the couple-stress theory for perfectly elastic, centrosymmetric-isotropic materials. It was pointed out by the authors that the new material constant ℓ, which has the dimension of length and which embodies the essential difference between analogous equations or solutions with and without couple stresses, is presumably small in comparison with bodily dimensions and wave lengths normally encountered, as there appears to be no conclusive experimental evidence of its existence for such materials as metals and polymers.

The theory discussed by Mindlin and Tiersten,[2.4] represents a rather special generalization of the classical theory of elasticity. In the early sixties a number of other such special cases were considered. It is not the purpose of this Section to present a survey of generalized theories for continua. For a rather general discussion we refer to the monographs by Stojanović, [2.5, 2.6].

Inspection of the system of governing equations derived in Section 1.2 suggests that the concepts and theories of a Cosserat continuum and its generalizations have applicability to the description of the mechanical behavior of directionally reinforced composites. Indeed, it was shown in Ref. 2. that a continuum theory for a laminated medium can be evolved which bears strong resemblance to Mindlin's theory of elasticity with microstructure (TEMS), see Ref. [2.8], or in its simpler version to the continuum discussed by Mindlin and Triesten, [2.4]. What is particularly noteworthy, however, is that the nonclassical material constants were evaluated as functions of the geometric layout and of the classical constants of the two constituent homogeneous materials of the composite.

The system of governing equations presented in the previous section is clearly more complicated than the system of equations for the theory of elasticity with microstructure which was stated in Ref. 2.8 . The most noticeable difference is the appearance of two sets of microdeformations, which are connected by a constraint condition, whereas TEMS contains only one set of microdeformations.

1.4. Propagation of Plane Harmonic Waves

As a check on the theory of Section 1.2. the dispersion equations for plane harmonic waves obtained from the displacement equations of motion (1.27)-(1.38) may be compared with the exact results of Section 4 of Part I. Thus, we may consider displacement components of the form

$$u_j = Ad_j \exp\left[ik(x_m p_m - ct)\right] \tag{1.39}$$

In Eq. (1.39) k is the wave number and c is the phase velocity.

As an example we will consider transverse waves propagating in the direction of the layering. For waves of this type the field variables are of the forms

$$(1.40)\quad \left(u_2\,,\psi_{21}^{(f)},\psi_{21}^{(m)},\Sigma_{21}\right) = \left(U_2\,,\Psi_{21}^{(f)},\Psi_{21}^{(m)},\Sigma_{21}\right)\exp\,[ik(x_1-ct)]$$

All other field variables vanish identically. The substitution of the expressions (1.40) in the displacement equations of motion yields a system of four homogeneous equations for U_2, $\Psi_{21}^{(f)}$, $\Psi_{21}^{(m)}$ and Σ_{21}. The dispersion equation is obtained from the requirement that the determinant of the coefficients must vanish. The dispersion equation is stated in detail in Ref. 2.1 . The equation yields curves relating the phase velocity and the wavenumber. For various values of the material constants dispersion curves are shown in Fig. 2.3. In this figure we have used

$$\eta = 0.8,\quad \rho_f/\rho_m = 3,$$

while the Poisson's ratios are

$$\nu_f = 0.3,\quad \nu_m = 0.35$$

Figure 2.3 shows the dimensionless phase velocity

$$\beta = c/(\mu_m/\rho_m)^{\frac{1}{2}}$$

versus the dimensionless wavenumber

$$\xi = kd_f\ ,$$

for various values of

$$\gamma = \mu_f/\mu_m$$

It is noted that the agreement with the exact theory, which was discussed in Section 4 of Part I, is very satisfactory. It can be shown that the results at $\xi = 0$ agree with the effective modulus theory.

Comparisons for other type of wave motions can be found in Ref. 2.1 .

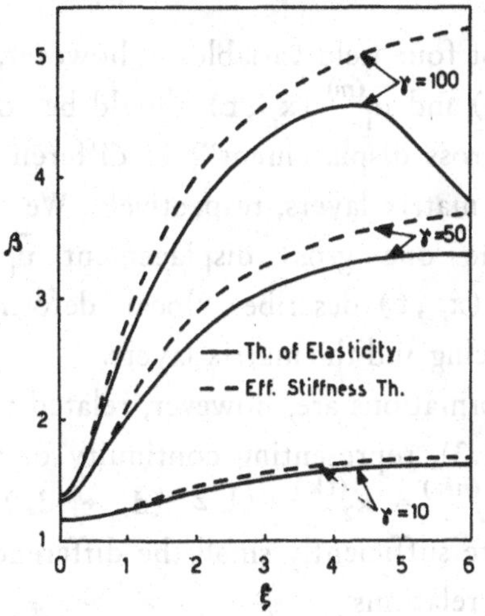

Fig. 2.3. Phase velocity versus wavenumber for transverse waves

1.5. An Alternate Way to Construct a Homogeneous Continuum Theory for a Laminated Medium

For constituent materials that are not elastic the approach which is based on the construction of strain and kinetic energy densities and the subsequent use of Hamilton's principle encounters difficulties. We present, therefore, also an alternate way of deriving the governing equations, which has the advantage that it can be used for other than elastic constitutive behavior of the individual layers.

Referring to Fig. 2.1 and Fig. 2.2 we take as point of departure the same displacement expansions as in Section 1.2 for the kth pair of layers. These expansions are given by Eqs. (1.1) and (1.2). The requirement that the displacements are continuous across the interfaces yields Eq. (1.3). The motion of the reinforcing layers and the matrix layers is thus described by the field variables $\bar{u}_i^{(fk)}$, $\psi_{2i}^{(fk)}$, and $\bar{u}_i^{(mk)}$, $\psi_{2i}^{(mk)}$, respectively. These field variables are defined only in discrete parallel planes $x_2 = x_2^{(fk)}$ and $x_2 = x_2^{(mk)}$, respectively. As a first step toward the construction of a homogeneous continuum model for the laminated medium we now introduce fields that are continuous functions of x_2 and whose values at $x_2 = x_2^{(fk)}$ and $x_2 = x_2^{(mk)}$ coincide with the actual field variables in the midplanes of the layers. The step is indicated symbolically by writing $\bar{u}_i^{(m)}(x_i, t)$ rather than $\bar{u}_i^{(mk)}(x_1, x_2^{(mk)}, x_3, t)$, etc. In this manner, four continuous fields are introduced:

$$\bar{u}_i^{(f)}(x_i, t),\quad \bar{u}_i^{(m)}(x_i, t),\ \psi_{2i}^{(f)}(x_i, t) \text{ and } \psi_{2i}^{(m)}(x_i, t).$$

The number of four field variables is, however, reduced to three by the argument that $\bar{u}_i^{(f)}(x_i, t)$ and $\bar{u}_i^{(m)}(x_i, t)$ should be considered as representing the same function (the "gross displacement") in different regions, namely, in the reinforcing layers and the matrix layers, respectively. We thus replace $\bar{u}_i^{(f)}(x_i, t)$ and $\bar{u}_i^{(m)}(x_i, t)$ by the one gross displacement $\bar{u}_i(x_i, t)$. The functions $\psi_{2i}^{(f)}(x_i, t)$ and $\psi_{2i}^{(m)}(x_i, t)$ describe "local deformations" which remain distinguished in the reinforcing and the matrix layers.

The local deformations are, however, related to the gross displacement through the conditions (1.3), representing continuity of the displacement at the interfaces. Noting that $x_2^{(mk)} = x_2^{(fk)} + (1/2)(d_m + d_f)$, and assuming that the thicknesses of the layers are sufficiently small the difference relations (1.3) can be replaced by the differential relations

$$\partial_2 u_i(x_i, t) = \eta\psi_{2i}^{(f)}(x_i, t) + (1-\eta)\psi_{2i}^{(m)}(x_i, t), \tag{1.41}$$

where η is defined by Eq. (1.18).

By the use of the small strain tensor, see Eq. (1.5), the assumed displacement distributions (1.1) and (1.2) are now used to compute the strains in the individual layers, where the differentiation in the x_2-direction should be with respect to the local coordinates $\bar{x}_2^{(f)}$ and $\bar{x}_2^{(m)}$. By introducing fields that are continuous not only in x_1 and x_3, but also in x_2, the transition from the discrete system of layers to the continuum model is effected, as discussed earlier in this section, and we find for the reinforcing layers

$$\begin{aligned}\epsilon_{\alpha\beta}^{(f)}\left(x_1, x_2, \bar{x}_2^{(f)}, x_3, t\right) &= \frac{1}{2}\left[\partial_\beta \bar{u}_\alpha(x_i, t) + \partial_\alpha \bar{u}_\beta(x_i, t)\right] \\ &+ \frac{1}{2}\bar{x}_2^{(f)}\left[\partial_\beta \psi_{2\alpha}^{(f)}(x_i, t) + \partial_\alpha \psi_{2\beta}^{(f)}(x_i, t)\right]\end{aligned} \tag{1.42}$$

$$\begin{aligned}\epsilon_{2\beta}^{(f)}\left(x_1, x_2, \bar{x}_2^{(f)}, x_3, t\right) = \epsilon_{\beta 2}^{(f)} &= \frac{1}{2}\left[\partial_\beta \bar{u}_2(x_i, t) + \psi_{2\beta}^{(f)}(x_i, t)\right] \\ &+ \frac{1}{2}\bar{x}_2^{(f)}\,\partial_\beta \psi_{22}^{(f)}(x_i, t)\end{aligned} \tag{1.43}$$

$$\epsilon_{22}^{(f)}\left(x_1, x_2, \bar{x}_2^{(f)}, x_3, t\right) = \psi_{22}^{(f)}(x_i, t). \tag{1.44}$$

In equations (1.42) and (1.43), α , β = 1,3. The summation convention must be invoked whenever the same subscript appears twice in the same term. Although (1.42)-(1.44) are fields whose values are defined as continuous functions of x_2, only the values in the discrete planes defined by $x = x_2^{(fk)}$ have physical relevance. The corresponding system for the matrix layers is obtained from (1.42)-(1.44) by replacing superscripts (f) by superscripts (m).

We now turn to the balance equations of linear momentum and moment of momentum for the homogeneous continuum model. These equations are derived directly from physical arguments based on momentum considerations of a pair of layers in the discrete system of layers.

For the kth reinforcing layer, the stress-equation of motion is

$$\partial_i \tau_{ij}^{(fk)} + \rho_f f_j^{(fk)} = \rho_f \ddot{u}_j^{(fk)} , \qquad |\bar{x}_2^{(f)}| \leq \frac{1}{2} d_f , \quad (1.45)$$

where $\tau_{ij}^{(fk)} = \tau_{ji}^{(fk)}$ is the stress tensor, ρ_f is the mass density, $f_j^{(fk)}$ is the body force, and a dot denotes a differentiation with respect to time. Similarly, for the kth matrix layer we write

$$\partial_i \tau_{ij}^{(mk)} + \rho_m f_j^{(mk)} = \rho_m \ddot{u}_j^{(mk)} , \qquad |\bar{x}_2^{(m)}| \leq \frac{1}{2} d_m . \quad (1.46)$$

To solve a boundary value problem for a layered medium rigorously, one has to find the solutions of equations (1.45) and (1.46) for each pair of layers and require continuity of the stresses and the displacements at the interfaces. In addition, the stresses and displacements must satisfy prescribed boundary conditions at the bounding surfaces of the laminated body. In the type of approximate analysis that is discussed in this Section it is argued, however, that if the characteristic length of the deformation is larger than the thicknesses of the laminations, it is not necessary to know the exact distributions of the stresses inside each layer to gain useful information on gross quantities such as displacements and frequencies. To compute approximations for such gross quantities it is sufficient to determine expressions for the stress resultant over a thickness and for the moment of the stresses over the thickness of a layer. This approach is consistent with the point of view that was taken earlier in assuming the displacement distributions (1.1) and (1.2). An analogous approach is followed in approximate plate theories. For the layered medium the stress resultants and the moments of the stresses over the thicknesses can then be used in conjunction with the balance equations of linear momentum to compute the interface stresses.

To derive the equations representing balance of linear momentum for the homogeneous continuum model, the stress equations of motion (1.45) and (1.46) are first integrated over their respective layer thicknesses and then added. As an illustration, we derive the equation of motion in the x_1-direction in some detail. This derivation involves several integrations over individual layer thicknesses, and for simplicity we introduce the following notation to indicate integration over the kth reinforcing layer of a function $g^{(fk)}(x_1, x_2^{(fk)}, \bar{x}_2^{(f)}, x_3, t)$:

$$\text{(1.47)} \quad Int^{(fk)}[g^{(fk)}] = \int_{-\frac{1}{2}d_f}^{\frac{1}{2}d_f} g^{(fk)}\left(x_1, x_2^{(fk)}, \bar{x}_2^{(f)}, x_3, t\right) d\bar{x}_2^{(f)} .$$

Similary, integration of $g^{(mk)}(x_1, x_2^{(mk)}, \bar{x}_2^{(m)} x_3, t)$ over the kth matrix layer is indicated by

$$\text{(1.48)} \quad Int^{(mk)}[g^{(mk)}] = \int_{-\frac{1}{2}d_m}^{\frac{1}{2}d_m} g^{(mk)}\left(x_1, x_2^{(mk)}, \bar{x}_2^{(m)}, x_3, t\right) d\bar{x}_2^{(m)} .$$

Then, from (1.45) and (1.46) we find

$$\frac{\partial \bar{\tau}_{11}^{(k)}}{\partial x_1} + \frac{\partial \bar{\tau}_{31}^{(k)}}{\partial x_3} + \frac{1}{d_f + d_m}\left\{\tau_{21}^{(mk)}\Big|\bar{x}_2^{(m)} = \frac{1}{2}d_m - \tau_{21}^{(fk)}\Big|\bar{x}_2^{(f)} = -\frac{1}{2}d_f\right\}$$

$$+ \rho \bar{f}_1^{(k)} = \bar{a}_1^{(k)} ,$$

(1.49)

where we have used the condition of continuity of the shear stress at the interface, and wherein

$$\text{(1.50)} \quad (d_f + d_m)\bar{\tau}_{11}^{(k)} = Int^{(fk)}\left[\tau_{11}^{(fk)}\right] + Int^{(mk)}\left[\tau_{11}^{(mk)}\right]$$

$$\text{(1.51)} \quad (d_f + d_m)\bar{\tau}_{31}^{(k)} = Int^{(fk)}\left[\tau_{31}^{(fk)}\right] + Int^{(mk)}\left[\tau_{31}^{(mk)}\right]$$

$$\text{(1.52)} \quad \rho(d_f + d_m)\bar{f}_1^{(k)} = Int^{(fk)}\left[\rho_f f_1^{(fk)}\right] + Int^{(mk)}\left[\rho_m f_1^{(mk)}\right]$$

and

$$\text{(1.53)} \quad (d_f + d_m)\bar{a}_1^{(k)} = Int^{(fk)}\left[\rho_f \ddot{u}_1^{(fk)}\right] + Int^{(mk)}\left[\rho_m \ddot{u}_1^{(mk)}\right] .$$

We now introduce the assumed displacement distributions (1.1) and (1.2) into (1.53) and carry out the integrations as indicated by (1.47) and (1.48), to obtain

$$(d_f + d_m)\,\bar{a}_1^{(k)} = d_f \rho_f \ddot{\bar{u}}_1^{(fk)}\left(x_1, x_2^{(fk)}, x_3, t\right) + d_m \rho_m \ddot{\bar{u}}_1^{(mk)}\left(x_1, x_2^{(mk)}, x_3, t\right) \qquad (1.54)$$

It is noted from (1.50) that $\bar{\tau}_{11}^{(k)}$ is an average stress over the kth pair of layers, which contains both $x_2^{(fk)}$ and $x_2^{(mk)}$ as parameters. To make the transition from the system of discrete layers to the homogeneous continuum model, we assume that $\bar{\tau}_{11}^{(k)}$ is the value at a position x_2 within the kth pair of layers of a field $\tau_{11}(x_i, t)$, which is continuous in x_2. The same argument is used for $\bar{\tau}_{31}^{(k)}$ and $\bar{f}_1^{(k)}$, and also for $\bar{a}_1^{(k)}$, where in the latter case $\bar{u}_1^{(fk)}$ and $\bar{u}_1^{(mk)}$, as previously, are considered as representing the same function, i.e., the gross displacement $\bar{u}_i(x_i, t)$, at different locations. If, moreover, the difference between $\ddot{\bar{u}}_i(x_i, t)$ at $x_2 = x_2^{(fk)}$ and $x_2 = x_2^{(mk)}$ is neglected, $\bar{a}_1^{(k)}$ may be replaced by $\bar{a}_1$, where

$$\bar{a}_1 = \rho \ddot{\bar{u}}_1(x_i, t), \qquad (1.55)$$

where the effective mass density ρ is defined as

$$\rho = \eta \rho_f + (1-\eta)\rho_m . \qquad (1.56)$$

For the remaining term in (1.49) we argue that the difference between the stresses at two interfaces, divided by the distance between the interfaces, can be represented by the derivative in the x_2-direction of a continuous function $\Sigma_{21}(x_i, t)$, i.e.,

$$\frac{1}{d_f + d_m}\left\{ \tau_{21}^{(mk)}\Big|_{\bar{x}_2^m = \frac{1}{2} d_m} - \tau_{21}^{(fk)}\Big|_{\bar{x}_2^f = -\frac{1}{2} d_f} \right\} \simeq \frac{\partial \Sigma_{21}}{\partial x_2} = \partial_2 \Sigma_{21}. \qquad (1.57)$$

The equation of motion for the homogeneous continuum model may then be written

$$\frac{\partial \bar{\tau}_{11}}{\partial x_1} + \frac{\partial \bar{\tau}_{31}}{\partial x_3} + \frac{\partial \Sigma_{21}}{\partial x_2} + \rho \bar{f}_1 = \rho \ddot{\bar{u}}_1 . \qquad (1.58)$$

The equations of motion in the x_2- and x_3-directions can be derived in a similar manner, and we may thus write

$$\partial_\alpha \bar{\tau}_{\alpha j} + \partial_2 \Sigma_{2j} + \rho \bar{f}_j = \rho \ddot{\bar{u}}_j , \tag{1.59}$$

where $\alpha = 1,3$ and $j = 1,2,3$. The stress components $\bar{\tau}_{\alpha j}$ and Σ_{2j} are referred to as composite stresses and interface stresses, respectively.

Returning to the system of discrete layers, we consider the kth reinforcing layer and we write the equation for balance of angular momentum of an element of the layer about the x_3-axis, by multiplying (1.45) by $\bar{x}_2^f$ and integrating over the thickness of the layer. The result is

$$\frac{\partial m_{121}^{(fk)}}{\partial x_1} + \frac{\partial m_{321}^{(fk)}}{\partial x_3} + \frac{1}{d_f} \mathrm{Int}^{(fk)}\left[\bar{x}_2^{(f)} \frac{\partial \tau_{21}^{(fk)}}{\partial \bar{x}_2^{(f)}}\right] + \rho_f \ell_{21}^{(fk)} = \omega_1^{(fk)} \tag{1.60}$$

wherein

$$d_f m_{121}^{(fk)} = \mathrm{Int}^{(fk)}\left[\bar{x}_2^{(f)} \tau_{11}^{(fk)}\right] \tag{1.61}$$

$$d_f m_{321}^{(fk)} = \mathrm{Int}^{(fk)}\left[\bar{x}_2^{(f)} \tau_{31}^{(fk)}\right] \tag{1.62}$$

$$d_f \rho_f \ell_{21}^{(fk)} = \mathrm{Int}^{(fk)}\left[\bar{x}_2^{(f)} \rho_f f_1^{(fk)}\right] \tag{1.63}$$

$$d_f \omega_1^{(fk)} = \mathrm{Int}^{(fk)}\left[\bar{x}_2^{(f)} \rho_f \ddot{u}_1^{(fk)}\right] . \tag{1.64}$$

The stress moments $m_{121}^{(fk)}$ and $m_{321}^{(fk)}$ are termed double stresses, where the first subscript designates the normal to the surface across which the component acts, the second subscript gives the orientation of the lever arm between the forces, and the third subscript gives the orientation of the forces. It is clear that $\ell_{21}^{(fk)}$ designates a body double force.

After an integration by parts the integral in (1.60) can be rewritten as

$$\frac{1}{d_f} \mathrm{Int}^{(fk)}\left[\bar{x}_2^{(f)} \frac{\partial \tau_{21}^{(fk)}}{\partial \bar{x}_2^{(f)}}\right] = \frac{1}{2}\left\{ \tau_{21}^{(fk)} \Big|_{\bar{x}_2^{(f)} = \frac{1}{2} d_f} + \tau_{21}^{(fk)} \Big|_{\bar{x}_2^{(f)} = -\frac{1}{2} d_f} \right\} - \bar{\tau}_{21}^{(fk)} ,$$

where

$$d_f \bar{\tau}_{21}^{(fk)} = \mathrm{Int}^{(fk)}\left[\tau_{21}^{(fk)}\right] . \tag{1.65}$$

If the assumed displacement distribution (1.1) is substituted in (1.64), we obtain

$$\omega_1^{(fk)} = I_f \ddot{\psi}_{21}^{(fk)} ,$$

where

$$I_f = \frac{1}{12} \rho_f d_f^2 . \qquad (1.66)$$

The transition from the system of discrete layers to the homogeneous continuum model is effected in exactly the same manner as for the equations of linear momentum. Thus we introduce the fields of double stresses $m_{\alpha 2j}$ (x_i, t), layer stresses $\bar{\tau}_{2j}^{(f)}$ (x_i, t), and interface stresses Σ_{2j} (x_i, t). For the homogeneous continuum model the equation for angular momentum then becomes

$$\frac{\partial m_{121}^{(f)}}{\partial x_1} + \frac{\partial m_{321}^{(f)}}{\partial x_3} + \Sigma_{21} - \bar{\tau}_{21}^{(f)} + \rho_f \ell_{21}^{(f)} = I_f \ddot{\psi}_{21}^{(f)} . \qquad (1.67)$$

In equation (1.67) we have used

$$\frac{1}{2}\left\{ \tau_{21}^{(fk)} \Big|_{\bar{x}_2^{(f)} = \frac{1}{2} d_f} + \tau_{21}^{(fk)} \Big|_{\bar{x}_2^{(f)} = -\frac{1}{2} d_f} \right\} \simeq \Sigma_{21} \Big|_{x_2 = x_2^{(fk)}} .$$

Angular momentum for the kth matrix layer can be treated in exactly the same manner. In fact, the corresponding equations can be obtained from (1.60) and (1.67) by replacing subscripts and superscripts f by subscripts and superscripts m.

Angular momentum about the other axes yields an analogous set of equations, and the equations for angular momentum of the homogeneous continuum model may thus be summarized as

$$\partial_\alpha m_{\alpha 2j}^{(f)} + \Sigma_{2j} - \bar{\tau}_{2j}^{(f)} + \rho_f \ell_{2j}^{(f)} = I_f \ddot{\psi}_{2j}^{(f)} \qquad (1.68)$$

$$\partial_\alpha m_{\alpha 2j}^{(m)} + \Sigma_{2j} - \bar{\tau}_{2j}^{(m)} + \rho_m \ell_{2j}^{(m)} = I_m \ddot{\psi}_{2j}^{(m)} \qquad (1.69)$$

where $\alpha = 1,3$, and $j = 1,2,3$.

If the interface stresses Σ_{2j} are eliminated from equations (1.59), (1.58), and (1.69), and if the constraint conditions (1.41) are introduced, the resulting system of equations, with appropriate redefinition of the stresses and double stresses, reduces to equations (57) and (58) of Ref. 2,2 . In this section, the interface stresses are considered unknown dynamical quantities which are to be determined from the balance equations.

Equations (1.59), (1.68), and (1.69) are the equations of motion for the homogeneous continuum model for the laminated composite. The form of the equations suggests the specification on an element of a bounding surface defined by the surface normal $\underset{\sim}{n}$, of a surface traction $T_{(\underset{\sim}{n})j}$ and surface moments $M^{(f)}_{(\underset{\sim}{n})2j}$ and $M^{(m)}_{(\underset{\sim}{n})2j}$. The external surface tractions and moments are related to the internal stresses and double stresses by

$$T_{(\underset{\sim}{n})j} = \bar{\tau}_{\alpha j}\, n_\alpha + \Sigma_{2j}\, n_2 \tag{1.70}$$

$$M^{(f)}_{(\underset{\sim}{n})2j} = m^{(f)}_{\alpha 2j}\, n_\alpha \tag{1.71}$$

$$M^{(m)}_{(\underset{\sim}{n})2j} = m^{(m)}_{\alpha 2j}\, n_\alpha \;. \tag{1.72}$$

As shown in [2.2], the relation between the surface tractions and moments and the actual distributions of simple tractions in the discrete system of layers can be determined quite easily for surfaces for which $n_2 \equiv 0$. It is possible to present analogous expressions for arbitrary orientation of the surface normal. From the physical point of view, conceptual difficulties appear, however, if the component in the x_1- and x_3-directions of the surface normal become very small. In fact, for $n_1 = n_3 \equiv 0$, the boundary conditions do not distinguish between a bounding surface in a matrix or a reinforcing layer. For the computation of gross dynamic quantities such as frequencies, and for a body with a sufficiently large number of layers, this may not be a serious drawback. The theory presented here is, however, not valid if the form of the physical boundary conditions on planes making small angles with the $x_1 x_3$-plane would suggest large variations of the stress and strain distributions across a layer thickness.

To complete the theory, constitutive equations are formulated for the composite stress $\bar{\tau}_{\alpha j}$, the double stresses $m^{(f)}_{\alpha 2j}$ and $m^{(m)}_{\alpha 2j}$, and the layer stresses $\bar{\tau}^{(f)}_{2j}$ and $\bar{\tau}^{(m)}_{2j}$. We start again with the constitutive equations for the kth reinforcing layer, which for an anisotropic linearly elastic solid may be written in the form

$$\tau^{(fk)}_{ij} = C^{(f)}_{ij\ell m}\, \epsilon^{(fk)}_{\ell m} \;, \tag{1.73}$$

where the elastic constants satisfy the symmetry conditions

$$C_{ij\ell m} = C_{ji\ell m} = C_{ijm\ell} = C_{\ell mij} \; . \qquad (1.74)$$

A completely analogous set of equations may be written for the kth matrix layer.

By computing the strains from (1.1) and substituting the results into (1.73), we may rewrite the latter equation as

$$\tau_{ij}^{(fk)}\left(x_1, x_2^{(fk)}, \bar{x}_2^{(f)}, x_3, t\right) = C_{ij\beta m}^{(f)} \partial_\beta \bar{u}_m^{(fk)} + C_{ij2m}^{(f)} \psi_{2m}^{(fk)} + \bar{x}_2^{(f)} C_{ij\beta m}^{(f)} \partial_\beta \psi_{2m}^{(fk)} , \qquad (1.75)$$

where $\beta = 1,3$, and $i,j,m = 1,2,3$. Again, similar equations can be written for the kth matrix layer. The equations for the stresses are now substituted in (1.50), (1.51), (1.64), to yield the average stresses $\bar{\tau}_{11}^{(k)}$, etc. The subsequent transition from the discrete system to the continuum model, as discussed earlier in this section, gives the composite stress $\bar{\tau}_{\alpha j}(x_i, t)$ as

$$\bar{\tau}_{\alpha j}(x_i, t) = C_{\alpha j\beta m} \partial_\beta \bar{u}_m + \eta C_{\alpha j2m}^{(f)} \psi_{2m}^{(f)} + (1-\eta) C_{\alpha j2m} \psi_{2m}^{(m)} , \qquad (1.76)$$

where we have introduced the "gross elastic constants"

$$C_{\alpha j\beta m} = \eta C_{\alpha j\beta m}^{(f)} + (1-\eta) C_{\alpha j\beta m}^{(m)} \qquad (1.77)$$

and $\alpha, \beta = 1,3$. The remaining constitutive relations are obtained in a completely analogous manner. Without further discussion we write the equations for the double stress $m_{\alpha 2j}^{(f)}(x_i, t)$ as

$$m_{\alpha 2j}^{(f)} = \frac{1}{12} d_f^2 C_{\alpha j\ell\beta}^{(f)} \partial_\beta \psi_{2\ell}^{(f)} \; . \qquad (1.78)$$

The corresponding equation for $m_{\alpha 2j}^{(m)}(x_i, t)$ can be obtained directly from (1.78). The layer stresses $\bar{\tau}_{2j}^{(f)}(x_i, t)$ may be written

$$\bar{\tau}_{2j}^{(f)} = C_{2j\ell\beta}^{(f)} \partial_\beta \bar{u}_\ell + C_{2j\ell 2}^{(f)} \psi_{2\ell}^{(f)} \qquad (1.79)$$

with a corresponding equation for $\bar{\tau}_{2j}^{(m)}(x_i, t)$. In equation (1.78) and (1.79), $\alpha, \beta = 1,3$, and the summation convention must be invoked. Equations (1.75)-(1.79) are the constitutive relations for the homogeneous continuum model for the laminated medium composed of anisotropic linearly elastic layers. The special case of isotropic layers will be presented next.

As a concluding comment, it is noted that the approach presented in this section can easily be extended to include the effects of viscoelasticity and temperature variation. For details we refer to Ref. 2.9 . An extension to large deformations was presented in Ref. 2.10 .

1.6. Summary of Equations for Isotropic Layers

In the previous section we have formulated a theory to describe the dynamic behavior of a laminated medium. The deformation is described by gross displacements $\bar{u}_i\ (x_j\ ,t)$, and local deformations $\psi_{2i}^{(f)}\ (x_j\ ,t)$, and $\psi_{2i}^{(m)}\ (x_j\ ,t)$, which are related by a constraint condition. The governing equations are balance laws and constitutive relations.

For an isotropic elastic solid the constants introduced in (1.73) have the special forms

$$C^{(f)}_{ij\ell m} = \lambda_f \delta_{ij}\delta_{\ell m} + \mu_f(\delta_{i\ell}\delta_{jm} + \delta_{im}\delta_{j\ell}) \ , \tag{1.80}$$

where δ_{ij} is the Kronecker delta and λ_f and μ_f are Lamé's elastic constants. As a consequence, the constitutive equations (1.76)-(1.79) reduce to

$$\bar{\tau}_{\alpha\beta}(x_i\ ,t) = \left\{\lambda\partial_\gamma\bar{u}_\gamma + \eta\lambda_f\psi_{22}^{(f)} + (1-\eta)\lambda_m\psi_{22}^{(m)}\right\}\delta_{\alpha\beta} + \mu(\partial_\beta\bar{u}_\alpha + \partial_\alpha\bar{u}_\beta) \tag{1.81}$$

$$\bar{\tau}_{\alpha 2}(x_i\ ,t) = \mu\partial_\alpha\bar{u}_2 + \eta\,\mu_f\psi_{2\alpha}^{(f)} + (1-\eta)\mu_m\psi_{2\alpha}^{(m)} \tag{1.82}$$

$$m^{(f)}_{\alpha 2\beta} = \frac{1}{12}\, d_f^2\lambda_f\partial_\gamma\psi_{2\gamma}^{(f)}\delta_{\alpha\beta} + \frac{1}{12}\, d_f^2\mu_f\left\{\partial_\beta\psi_{2\alpha}^{(f)} + \partial_\alpha\psi_{2\beta}^{(f)}\right\} \tag{1.83}$$

$$m^{(f)}_{\alpha 22} = \frac{1}{12}\, d_f^2\mu_f\partial_\alpha\psi_{22}^{(f)} \tag{1.84}$$

$$\bar{\tau}^{(f)}_{2\beta} = \mu_f\partial_\beta\bar{u}_2 + \mu_f\psi_{2\beta}^{(f)} \tag{1.85}$$

$$\bar{\tau}^{(f)}_{22} = \lambda_f\partial_\gamma\bar{u}_\gamma + \lambda_f\psi_{22}^{(f)} + 2\mu_f\psi_{22}^{(f)} \tag{1.86}$$

In these equations, $\alpha,\beta,\gamma = 1,3$. The corresponding equations for $m^{(m)}_{\alpha 2j}$, $m^{(m)}_{\alpha 22}$, $\bar{\tau}^{(m)}_{2\beta}$, and $\bar{\tau}^{(m)}_{22}$ are directly obtained by replacing superscripts and subscripts f by superscripts and subscripts m. In equations (1.81)-(1.86) we have used

$$\lambda = \eta\lambda_f + (1-\eta)\lambda_m \tag{1.87}$$

$$\mu = \eta\mu_f + (1-\eta)\mu_m \; . \tag{1.88}$$

In the absence of body forces the equations of motion (1.59), (1.68) and (1.69) reduce to

$$\partial_\alpha \bar{\tau}_{\alpha j} + \partial_2 \Sigma_{2j} = \rho \ddot{\bar{u}}_j \tag{1.89}$$

$$\partial_\alpha m^{(f)}_{\alpha 2j} + \Sigma_{2j} - \bar{\tau}^{(f)}_{2j} = I_f \ddot{\psi}^{(f)}_{2j} \tag{1.90}$$

and

$$\partial_\alpha m^{(m)}_{\alpha 2j} + \Sigma_{2j} - \bar{\tau}^{(m)}_{2j} = I_m \ddot{\psi}^{(m)}_{2j} \tag{1.91}$$

The system of equations is completed by the continuity conditions (1.41) and the boundary conditions given by (1.70)-(1.72).

If the constitutive equations are substituted in the equations of motion, (1.89)-(1.91) yield the earlier derived equations (1.27)-(1.35).

1.7. Reduction to the Effective Modulus Theory

For a layered medium the effective modulus theory, whereby the layered medium is replaced by a homogeneous but anisotropic continuum whose elastic moduli and average mass density are expressed in terms of the dimensional parameters and the material constants of the constituents, was discussed in Part I. It can now easily be shown that the effective modulus theory is a special case of the more general continuum theory with microstructure which was discussed in the preceding sections. Here we demonstrate the reduction for the special case when the individual layers are homogeneous and isotropic, and the mechanical behavior of the laminated composite may be described by a transversely isotropic continuum with the axis of symmetry normal to the layering.

The reduction to the effective modulus theory can be achieved by neglecting all terms containing d_f^2 and d_m^2. From the constitutive equations (1.83) and (1.84) we then conclude

(1.92) $$m^{(f)}_{\alpha 2\beta} = m^{(f)}_{\alpha 22} = m^{(m)}_{\alpha 2\beta} = m^{(m)}_{\alpha 22} \equiv 0 \,.$$

The balance equations (1.90) and (1.91) then reduce to

(1.93) $$\Sigma_{2j} = \bar{\tau}^{(f)}_{2j}$$

and

(1.94) $$\Sigma_{2j} = \bar{\tau}^{(m)}_{2j}$$

Thus,

(1.95) $$\bar{\tau}^{(f)}_{2j} = \bar{\tau}^{(m)}_{2j} \,.$$

By employing the constitutive equations for $\bar{\tau}^{(f)}_{2j}$, as given by (1.85) and (1.86), as well as the corresponding equations for $\bar{\tau}^{(m)}_{2j}$, we obtain from (1.95) three expressions relating $\partial_j \bar{u}_2$, $\partial_2 \bar{u}_j$, $\psi^{(f)}_{2j}$ and $\psi^{(m)}_{2j}$, Three other relations between these quantities are given by (1.41), and we can thus solve for $\psi^{(f)}_{2j}$ and $\psi^{(m)}_{2j}$ in terms of $\partial_j \bar{u}_2$ and $\partial_2 \bar{u}_j$. The results are

(1.96) $$\psi^{(f)}_{2\alpha} = \frac{(1-\eta)[\mu_f - \mu_m]\partial_\alpha \bar{u}_2 + \mu_m \partial_2 \bar{u}_\alpha}{(1-\eta)\mu_f + \eta\,\mu_m}$$

(1.97) $$\psi^{(f)}_{22} = \frac{(1-\eta)[\lambda_m - \lambda_f]\partial_\gamma \bar{u}_\gamma + [\lambda_m + 2\mu_m]\partial_2 \bar{u}_2}{(1-\eta)[\lambda_f + 2\mu_f] + \eta[\lambda_m + 2\mu_m]}$$

(1.98) $$\psi^{(m)}_{2j} = \frac{\partial_2 \bar{u}_j - \eta\,\psi^{(f)}_{2j}}{1-\eta}$$

where $\alpha = 1,3$. If (1.96)-(1.98) are substituted into (1.81) we find expressions for $\bar{\tau}_{11}$, $\bar{\tau}_{13} = \bar{\tau}_{31}$, and $\bar{\tau}_{33}$. From (1.82) find expressions for $\bar{\tau}_{12}$ and $\bar{\tau}_{32}$, whereupon it can be checked by comparing (1.82) and (1.85) that

(1.99) $$\bar{\tau}_{12} = \bar{\tau}^{(f)}_{21} = \bar{\tau}^{(m)}_{21}$$

(1.100) $$\bar{\tau}_{32} = \bar{\tau}^{(f)}_{23} = \bar{\tau}^{(m)}_{23} \,.$$

Finally from (1.86) we obtain expressions for $\bar{\tau}_{22}^{(f)} = \bar{\tau}_{22}^{(m)}$, which we relabel as the composite stress $\bar{\tau}_{22}$:

$$\bar{\tau}_{22} = \bar{\tau}_{22}^{(f)} = \bar{\tau}_{22}^{(m)} . \tag{1.101}$$

The stress-strain relations for τ_{ij} may then be summarized as

$$\bar{\tau}_{ij} = A^*_{ijk\ell}\, \epsilon_{k\ell} , \tag{1.102}$$

wherein

$$A^*_{1111} = A^*_{3333} = [(\lambda_f + 2\mu_f)(\lambda_m + 2\mu_m) + 4\eta(1-\eta)(\mu_f - \mu_m)(\lambda_f + \mu_f - \lambda_m - \mu_m)]/D$$

$$A^*_{1122} = A^*_{2211} = A^*_{2233} = A^*_{3322} = [\eta\lambda_f(\lambda_m + 2\mu_m) + (1-\eta)\lambda_m(\lambda_f + 2\mu_f)]/D$$

$$A^*_{1133} = A^*_{3311} = \{2[\eta\lambda_f + (1-\eta)\lambda_m][(1-\eta)\mu_f + \eta\mu_m] + \lambda_f\lambda_m\}/D$$

$$A^*_{2222} = (\lambda_f + 2\mu_f)(\lambda_m + 2\mu_m)/D .$$

$$A^*_{1212} = A^*_{1221} = A^*_{2121} = A^*_{2112} = A^*_{3232} = A^*_{2323} = A^*_{3223} = A^*_{2332} = 2\mu_f\mu_m/[(1-\eta)\mu_f + \eta\mu_m]$$

$$A^*_{1313} = A^*_{1331} = A^*_{3113} = A^*_{3131} = A^*_{1111} = A^*_{1133}$$

$$D = (1-\eta)(\lambda_f + 2\mu_f) + \eta(\lambda_m + 2\mu_m) .$$

The constants $A^*_{ijk\ell}$ that are not explicitly defined are identically zero. The five effective elastic constants agree with the expressions that were derived earlier.

The stress-equations of motion for the effective modulus theory are obtained by substituting (1.93), (1.94), (1.99), (1.100) and (1.101) in (1.89). We obtain

$$\partial_i \bar{\tau}_{ij} = \rho \ddot{\bar{u}}_j . \tag{1.103}$$

The displacement equations of motion are obtained by substituting (1.102) into (1.103).

CHAPTER 2

A THEORY OF ELASTICITY WITH MICROSTRUCTURE FOR FIBER-REINFORCED COMPOSITES

2.1. Introduction

In this Chapter the basic ideas of Chapter 1 are applied to derive a system of displacement equations of motion for a unidirectionally fiber-reinforced composite. The system of equations is derived in three stages. The first stage of the derivation involves certain assumptions and calculations within a representative cell of the actual fiber-reinforced composite. In particular it is assumed that the motion of a fiber and the neighboring matrix material can be described by linear expansions in a system of local coordinates. The kinematic variables that are introduced in the expansions are defined at the centerlines of the fibers only. On the basis of these expansions the strain energy and the kinetic energy in a representative cell are subsequently computed. The averages over the cell next yield energy densities which are defined at the centerline of the fiber. In the next stage of the derivation a transition is achieved from the system of discrete cells to a homogeneous continuum model in the manner discussed in Chapter 1. In the final stage Hamilton's principle in conjunction with certain continuity relations and the use of Lagrangian multipliers yields a set of displacement equations of motion.

In the last part of the Chapter the displacement equations of motion are employed to examine the propagation of transverse waves in the direction of the fibers.

The presentation of the material of this chapter follows the work of Achenbach and Sun, [2.11].

2.2. Governing Equations

We consider a fiber-reinforced composite consisting of unidirectional fibers embedded in a matrix material. It is assumed that the fibers are cylindrical rods of radius α arranged in rectangular arrays. The distance between the centerlines of the fibers are d_2 and d_3, in the x_2- and x_3-directions, respectively, as shown in Fig. 2.4. Each fiber is identified by two indices; the first index identifies the row and the second index identifies the column in which the fiber is located. The position of the centerline of fiber (k,ℓ) is defined by $x_2 = x_2^{\ell}$ and $x_3 = x_3^{k}$. The elastic constants of the high-modulus reinforcing fibers and the low-modulus matrix material are denoted by λ_f, μ_f and λ_m, μ_m, respectively.

To describe the displacement field the fiber-reinforced medium is divided into strips by the planes of structural symmetry of the composite, see Fig. 2.4. Each strip is of width d_2 and of height d_3, and each strip contains one fiber. We focus attention on the strip which contains fiber (k,ℓ). An element of unit length of this strip is labeled cell (k,ℓ). Next we define a system of local cylindrical coordinates, r, θ, x_1, as well as a system of local Cartesian coordinates $x_1, \bar{x}_2, \bar{x}_3$, see Fig. 2.5. Now, provided that the characteristic length of the deformation is sufficiently larger than either d_2 or d_3, the displacements in the cell can be approximated by expansions in terms of quantities which are defined at the centerline of the fiber, which is also the centerline of the cell. These expansions are analogous to the expansions used in rod theories. We write (in indicial notation $i = 1,2,3$),

in fiber (k,ℓ), $(r < \alpha)$:

$$u_i^{f(k,\ell)} = \bar{u}_i^{(k,\ell)} + r\cos\theta\, \psi_{2i}^{f(k,\ell)} + r\sin\theta\, \psi_{3i}^{f(k,\ell)} \tag{2.1}$$

in the matrix material of cell (k,ℓ), $(r > \alpha)$:

$$u_i^{m(k,\ell)} = \bar{u}_i^{(k,\ell)} + \alpha\cos\theta\, \psi_{2i}^{f(k,\ell)} + \alpha\sin\theta\, \psi_{3i}^{f(k,\ell)} + (r-\alpha)\cos\theta\, \psi_{2i}^{m(k,\ell)} + (r-\alpha)\sin\theta\, \psi_{3i}^{m(k,\ell)} \tag{2.2}$$

Thus, the displacement in the matrix is expressed as the displacement at the fiber-matrix interface plus additional terms which increase linearly with the distance from the interface. By expressing $u_i^{m(k,\ell)}$ in the form (2.2) the displacement satisfies the condition of continuity at the fiber-matrix interface. Equation (2.2) can also be written in the form

$$u_i^{m(k,\ell)} = \bar{u}_i^{(k,\ell)} + \alpha\cos\theta\left(\psi_{2i}^{f(k,\ell)} - \psi_{2i}^{m(k,\ell)}\right) + \alpha\sin\theta\left(\psi_{3i}^{f(k,\ell)} - \psi_{3i}^{m(k,\ell)}\right) + \bar{x}_2\,\psi_{2i}^{m(k,\ell)} + \bar{x}_3\,\psi_{3i}^{m(k,\ell)} \tag{2.3}$$

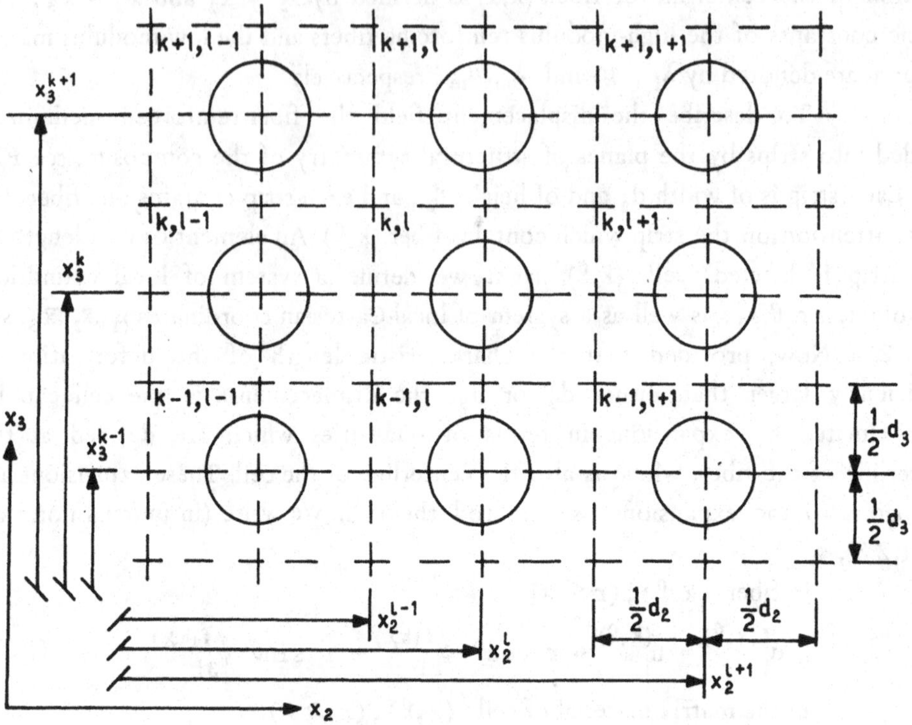

Fig. 2.4. Fiber-reinforced composite

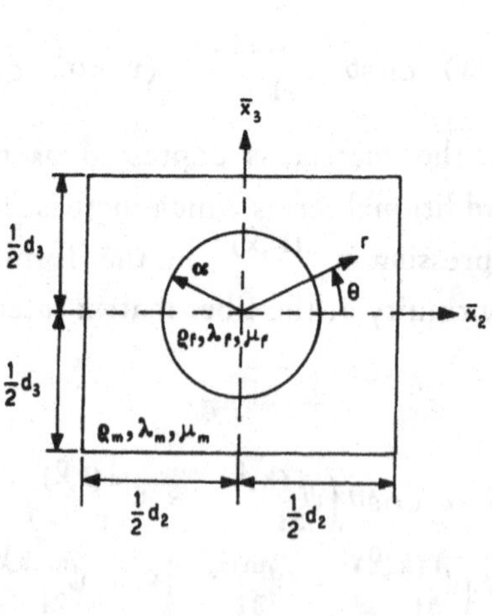

Fig. 2.5. Cell (k, ℓ) of the fiber-reinforced composite

The physical interpretation of the terms in Eqs. (2.1) and (2.2) clearly is that the "gross displacement" $\bar{u}_i^{(k,\ell)}$ is the displacement at the centerline (also the average displacement), while $\psi_{21}^{f(k,\ell)}$, $\psi_{23}^{f(k,\ell)}$, $\psi_{21}^{m(k,\ell)}$, and $\psi_{23}^{m(k,\ell)}$ represent thickness shear motions, and $\psi_{22}^{f(k,\ell)}$ and $\psi_{22}^{m(k,\ell)}$ represent thickness stretch motions. The field quantities and their dependence on the coordinates are summarized as

gross displacement	$\bar{u}_i^{(k,\ell)}\left(x_1, x_2^{\ell}, x_3^{k}, t\right)$
local fiber deformations	$\psi_{2i}^{f(k,\ell)}\left(x_1, x_2^{\ell}, x_3^{k}, t\right)$
	$\psi_{3i}^{f(k,\ell)}\left(x_1, x_2^{\ell}, x_3^{k}, t\right)$
local matrix deformations	$\psi_{2i}^{m(k,\ell)}\left(x_1, x_2^{\ell}, x_3^{k}, t\right)$
	$\psi_{3i}^{m(k,\ell)}\left(x_1, x_2^{\ell}, x_3^{k}, t\right)$

Note that within the actual fiber-reinforced composite the gross displacements and the local deformations are defined at discrete values of x_2 and x_3, but they are continuous functions of x_1 and t.

The displacements should be continuous at the interfaces between cell (k,ℓ) and the neighboring cells. It is, however, not possible to require point by point continuity. What can be done is to impose the condition that the average displacement is continuous at the interfaces of the cells. Thus, at the interface between cell (k,ℓ) and cell $(k+1,\ell)$ we require

$$\int_{-\frac{1}{2}d_2}^{\frac{1}{2}d_2}\left\{\left[u_i^{m(k+1,\ell)}\right]_{\bar{x}_3=-\frac{1}{2}d_3}-\left[u_i^{m(k,\ell)}\right]_{\bar{x}_3=\frac{1}{2}d_3}\right\}d\bar{x}_2=0 \tag{2.4}$$

Similarly, at the interface between cells (k,ℓ) and $(k,\ell+1)$ we require

$$\int_{-\frac{1}{2}d_3}^{\frac{1}{2}d_3}\left\{\left[u_i^{m(k,\ell+1)}\right]_{\bar{x}_2=-\frac{1}{2}d_2}-\left[u_i^{m(k,\ell)}\right]_{\bar{x}_2=\frac{1}{2}d_2}\right\}d\bar{x}_3=0 \tag{2.5}$$

Substituting Eq. (2.3) into (2.4) we obtain upon working out the integrals

$$\bar{u}_i^{(k+1,\ell)}-\bar{u}_i^{(k,\ell)}-\frac{\alpha d_3}{d_2}\ln\left[\frac{1+(1+\zeta^2)^{\frac{1}{2}}}{\zeta}\right]\left(\psi_{3i}^{f(k+1,\ell)}-\psi_{3i}^{m(k+1,\ell)}+\right.$$

$$(2.6) \qquad + \psi_{3i}^{f(k,\ell)} - \psi_{3i}^{m(k,\ell)}\Big) - \frac{1}{2}\, d_3\left(\psi_{3i}^{m(k+1,\ell)} + \psi_{3i}^{m(k,\ell)}\right) = 0$$

In Eq. (2.6) the ratio ζ is defined as

$$(2.7) \qquad \zeta = \frac{d_3}{d_2}$$

Similarly we obtain from Eq. (2.5)

$$\bar{u}_i^{(k,\ell+1)} - \bar{u}_i^{(k,\ell)} - \frac{\alpha d_2}{d_3}\, \ln\left[\zeta + (1+\zeta^2)^{\frac{1}{2}}\right]\Big(\psi_{2i}^{f(k,\ell+1)} - \psi_{2i}^{m(k,\ell+1)}$$

$$(2.8) \qquad + \psi_{2i}^{f(k,\ell)} - \psi_{2i}^{m(k,\ell)}\Big) - \frac{1}{2}\, d_2\left(\psi_{2i}^{m(k,\ell+1)} + \psi_{3i}^{m(k,\ell)}\right) = 0$$

The displacement expansions (2.1) and (2.3) can be used to compute the corresponding strains. Substituting Eq. (2.1) into the expression for the components of the small strain tensor,

$$\epsilon_{ij} = \frac{1}{2}\,(\partial_j u_i + \partial_i u_j),$$

in which

$$\partial_j u_i = \partial u_i / \partial x_j \quad ,$$

and where the differentiation in the x_2- and x_3-directions should be with respect to the local coordinates $\bar{x}_2$ and $\bar{x}_3$, we find

$$(2.9) \qquad \epsilon_{\gamma\gamma}^{f(k,\ell)} = \psi_{\gamma\gamma}^{f(k,\ell)} \qquad \text{(no summation)}$$

$$(2.10) \qquad \epsilon_{23}^{f(k,\ell)} = \frac{1}{2}\left(\psi_{23}^{f(k,\ell)} + \psi_{32}^{f(k,\ell)}\right)$$

$$(2.11) \qquad \epsilon_{11}^{f(k,\ell)} = \partial_1 \bar{u}_1^{(k,\ell)} + \bar{x}_2\, \partial_1 \psi_{21}^{f(k,\ell)} + \bar{x}_3\, \partial_1 \psi_{31}^{f(k,\ell)}$$

$$(2.12) \qquad \epsilon_{1\gamma}^{f(k,\ell)} = \epsilon_{\gamma 1}^{f(k,\ell)} = \frac{1}{2}\left(\partial_1 \bar{u}_\gamma^{(k,\ell)} + \bar{x}_2\, \partial_1 \psi_{2\gamma}^{f(k,\ell)} + \bar{x}_3\, \partial_1 \psi_{3\gamma}^{f(k,\ell)} + \psi_{\gamma 1}^{f(k,\ell)}\right)$$

In Eqs. (2.9)-(2.12) the Greek index γ can assume the values 2 or 3 only. For the matrix material we find from Eqs. (2.3)

$$\epsilon_{\gamma\gamma}^{m(k,\ell)} = \psi_{\gamma\gamma}^{m(k,\ell)} + \left(\psi_{2\gamma}^{f(k,\ell)} - \psi_{2\gamma}^{m(k,\ell)}\right)\alpha \frac{\partial \cos\theta}{\partial \bar{x}_\gamma} + \left(\psi_{3\gamma}^{f(k,\ell)} - \psi_{3\gamma}^{m(k,\ell)}\right)\alpha \frac{\partial \sin\theta}{\partial \bar{x}_\gamma} \tag{2.13}$$

$$\begin{aligned}\epsilon_{23}^{m(k,\ell)} = {} & \frac{1}{2}\left(\psi_{22}^{f(k,\ell)} - \psi_{22}^{m(k,\ell)}\right)\alpha \frac{\partial \cos\theta}{\partial \bar{x}_3} + \frac{1}{2}\left(\psi_{23}^{f(k,\ell)} - \psi_{23}^{m(k,\ell)}\right)\alpha \frac{\partial \cos\theta}{\partial \bar{x}_2} \\ & + \frac{1}{2}\left(\psi_{32}^{f(k,\ell)} - \psi_{32}^{m(k,\ell)}\right)\alpha \frac{\partial \sin\theta}{\partial \bar{x}_3} + \frac{1}{2}\left(\psi_{33}^{f(k,\ell)} - \psi_{33}^{m(k,\ell)}\right)\alpha \frac{\partial \sin\theta}{\partial \bar{x}_2} \\ & + \frac{1}{2}\left(\psi_{23}^{m(k,\ell)} + \psi_{32}^{m(k,\ell)}\right)\end{aligned} \tag{2.14}$$

$$\begin{aligned}\epsilon_{11}^{m(k,\ell)} = {} & \partial_1 \bar{u}_1^{(k,\ell)} + \alpha \cos\theta\left(\partial_1 \psi_{21}^{f(k,\ell)} - \partial_1 \psi_{21}^{m(k,\ell)}\right) \\ & + \alpha \sin\theta\left(\partial_1 \psi_{31}^{f(k,\ell)} - \partial_1 \psi_{31}^{m(k,\ell)}\right) + \bar{x}_2 \partial_1 \psi_{21}^{m(k,\ell)} + \bar{x}_3 \partial_1 \psi_{31}^{m(k,\ell)}\end{aligned} \tag{2.15}$$

$$\begin{aligned}\epsilon_{1\gamma}^{m(k,\ell)} = {} & \frac{1}{2}\Big[\partial_1 \bar{u}_\gamma^{(k,\ell)} + \alpha \cos\theta\left(\partial_1 \psi_{2\gamma}^{f(k,\ell)} - \partial_1 \psi_{2\gamma}^{m(k,\ell)}\right) + \alpha \sin\theta\Big(\partial_1 \psi_{3\gamma}^{f(k,\ell)} \\ & - \partial_1 \psi_{3\gamma}^{m(k,\ell)}\Big) + \bar{x}_2 \partial_1 \psi_{2\gamma}^{m(k,\ell)} + \bar{x}_3 \partial_1 \psi_{3\gamma}^{m(k,\ell)}\Big] + \frac{1}{2}\Big[\Big(\psi_{21}^{f(k,\ell)} \\ & - \psi_{21}^{m(k,\ell)}\Big)\alpha \frac{\partial \cos\theta}{\partial \bar{x}_\gamma} + \left(\psi_{31}^{f(k,\ell)} - \psi_{31}^{m(k,\ell)}\right)\alpha \frac{\partial \sin\theta}{\partial \bar{x}_\gamma} + \psi_{\gamma 1}^{m(k,\ell)}\Big]\end{aligned} \tag{2.16}$$

The displacement expansions (2.1)-(2.3) can also be used to compute particle velocities. For the kinetic energy stored in the fiber element of cell (k,ℓ) we find

$$T^{f(k,\ell)} = \frac{1}{2}\rho_f \iint_{A_f} \sum_{i=1}^{3} \left(\dot{u}_i^{f(k,\ell)}\right)^2 dA_f \ , \tag{2.17}$$

where $A_f = \pi\alpha^2$ is the cross-sectional area of the fiber. By employing Eq. (2.1) we find

$$T^{f(k,\ell)} = \frac{1}{2}\rho_f \sum_{i=1}^{3} \left\{ A_f\left(\dot{\bar{u}}_i^{f(k,\ell)}\right)^2 + I_3^f\left(\dot{\psi}_{2i}^{f(k,\ell)}\right)^2 + I_2^f\left(\dot{\psi}_{3i}^{f(k,\ell)}\right)^2 \right\} , \tag{2.18}$$

where

$$I_2^f = \iint_{A_f} \bar{x}_3^2 \, dA_f \tag{2.19}$$

$$I_3^f = \iint_{A_f} \bar{x}_2^2 \, dA_f \tag{2.20}$$

The kinetic energy stored in the matrix material of element (k, ℓ) is

$$T^{m(k,\ell)} = \frac{1}{2}\rho_m \iint_{A_m} \sum_{i=1}^{3} \left(\dot{u}_i^{m(k,\ell)}\right)^2 dA_m \ , \tag{2.21}$$

where A_m is the cross-sectional area of the matrix material in cell (k,ℓ) i.e.

$$A_m = d_2 d_3 - \pi\alpha^2 \tag{2.22}$$

By employing Eqs. (2.2) we find

$$\begin{aligned} T^{m(k,\ell)} = \frac{1}{2}\rho_m \sum_{i=1}^{3} \Big\{ & A_m\left(\dot{\bar{u}}_i^{(k,\ell)}\right)^2 + J_1^m\left(\dot{\psi}_{2i}^{f(k,\ell)} - \dot{\psi}_{2i}^{m(k,\ell)}\right)^2 + J_2^m\left(\dot{\psi}_{3i}^{f(k,\ell)} \right. \\ & \left. - \dot{\psi}_{3i}^{m(k,\ell)}\right)^2 + I_3^m\left(\dot{\psi}_{2i}^{m(k,\ell)}\right)^2 + I_2^m\left(\dot{\psi}_{3i}^{m(k,\ell)}\right)^2 + 2J_6^m\left(\dot{\psi}_{2i}^{m(k,\ell)}\right) \\ & \times\left(\dot{\psi}_{2i}^{f(k,\ell)} - \dot{\psi}_{2i}^{m(k,\ell)}\right) + 2J_7^m\left(\dot{\psi}_{3i}^{m(k,\ell)}\right)\left(\dot{\psi}_{3i}^{f(k,\ell)} - \dot{\psi}_{3i}^{m(k,\ell)}\right) \Big\} \end{aligned} \tag{2.23}$$

The coeeficients in this expression, together with several other coefficients which appear later, are

$$I_2^m = \iint_{A_m} \bar{x}_3^2 \, dA_m \qquad (2.24)$$

$$I_3^m = \iint_{A_m} \bar{x}_2^2 \, dA_m \qquad (2.25)$$

$$J_1^m = \iint_{A_m} \alpha^2 \cos^2\theta \, dA_m \qquad (2.26)$$

$$J_2^m = \iint_{A_m} \alpha^2 \sin^2\theta \, dA_m \qquad (2.27)$$

$$J_3^m = \iint_{A_m} \frac{\alpha^2}{r^2} \cos^4\theta \, dA_m \qquad (2.28)$$

$$J_4^m = \iint_{A_m} \frac{\alpha^2}{r^2} \sin^4\theta \, dA_m \qquad (2.29)$$

$$J_5^m = \iint_{A_m} \frac{\alpha^2}{r^2} \sin^2\theta \cos^2\theta \, dA_m \qquad (2.30)$$

$$J_6^m = \iint_{A_m} \alpha \, r \cos^2\theta \, dA_m \qquad (2.31)$$

$$J_7^m = \iint_{A_m} \alpha \, r \sin^2\theta \, dA_m \qquad (2.32)$$

$$J_8^m = \iint_{A_m} \frac{\alpha}{r} \sin^2\theta \, dA_m \qquad (2.33)$$

$$J_9^m = \iint_{A_m} \frac{\alpha}{r} \cos^2\theta \, dA_m \qquad (2.34)$$

The total kinetic energy stored in cell (k,ℓ) is the sum of Eq. (2.18) and Eq. (2.23). The average over the volume of the cell is

$$T_{ave}^{(k,\ell)} = \frac{1}{d_2 d_3} \left(T^{f(k,\ell)} + T^{m(k,\ell)} \right) \qquad (2.35)$$

In an isotropic linearly elastic body the strain energy density can be written as

$$W = \frac{1}{2} (\lambda + 2\mu) \left(\epsilon_{11}^2 + \epsilon_{22}^2 + \epsilon_{33}^2 \right) + \lambda \left(\epsilon_{11} \epsilon_{22} + \epsilon_{11} \epsilon_{33} + \epsilon_{22} \epsilon_{33} \right) + 2\mu \left(\epsilon_{12}^2 + \epsilon_{23}^2 + \epsilon_{13}^2 \right) ,$$

where λ and μ are Lamé's elastic constants, and ϵ_{ij} are the components of the strain tensor.

Substitution of the strains (2.9)-(2.12) into W, and integration over the area A_f yields the strain energy $W^{f(k,\ell)}$ stored in the fiber-element of cell (k,ℓ). Substitution of Eqs. (2.13)-(2.16) into W and integration over A_m yields the strain energy $W^{m(k,\ell)}$ stored in the matrix material of cell (k,ℓ). The total strain energy averaged over the volume of cell (k,ℓ) yields

$$W_{ave}^{(k\ell)} = \frac{1}{d_2 d_3}\left(W^{f(k,\ell)} + W^{m(k,\ell)}\right) \tag{2.36}$$

The displacement distribution in the fiber-reinforced composite is now described by the field variables $\bar{u}_i^{(k,\ell)}$, $\psi_{2i}^{f(k,\ell)}$, $\psi_{3i}^{f(k,\ell)}$, $\psi_{2i}^{m(k,\ell)}$ and $\psi_{3i}^{m(k,\ell)}$. These variables are defined only on discrete lines $x_2 = x_2^\ell$ and $x_3 = x_3^k$. To obtain a continuum model we now introduce fields that are continuous in x_2 and x_3, and whose values at $x_2 = x_2^\ell$ and $x_3 = x_3^k$ coincide with the values of the actual field variables at the centerlines of the cells. The step is indicated by writing $\bar{u}_i(x_i, t)$ rather than $\bar{u}_i^{(k,\ell)}(x_1, x_2^\ell, x_3^k, t)$, etc. In this manner five continuous fields are introduced:

gross displacements $\bar{u}_i(x_j, t)$

local deformations $\psi_{2i}^f(x_j, t)$ and $\psi_{3i}^f(x_j, t)$

$\psi_{2i}^m(x_j, t)$ and $\psi_{3i}^m(x_j, t)$

As a direct implication of the foregoing step we can also consider a strain energy density $W(x_i,t)$ and a kinetic energy density $T(x_i,t)$ which are continuous functions of x_i and t, and whose values at $x_2 = x_2^k$ and $x_3 = x_3^\ell$ agree with $W_{ave}^{(k\ell)}$ and $T_{ave}^{(k\ell)}$. The kinetic energy density follows from Eqs. (2.35), (2.17) and (2.23) as

$$A_c T = \frac{1}{2}\rho_f \sum_{i=1}^{3}\left\{A_f(\dot{\bar{u}}_i)^2 + I_3^f\left(\dot{\psi}_{2i}^f\right)^2 + I_2^f\left(\dot{\psi}_{3i}^f\right)^2\right\} + \frac{1}{2}\rho_m \sum_{i=1}^{3}\left\{A_m(\dot{\bar{u}}_i)^2\right.$$

$$\left. + J_1^m\left(\dot{\psi}_{2i}^f - \dot{\psi}_{2i}^m\right)^2 + J_2^m\left(\dot{\psi}_{3i}^f - \dot{\psi}_{3i}^m\right)^2 + I_3^m\left(\dot{\psi}_{2i}^m\right)^2 + I_2^m\left(\dot{\psi}_{3i}^m\right)^2\right.$$

$$+ 2J_6^m\left(\dot{\psi}^m_{2i}\right)\left(\dot{\psi}^f_{2i} - \dot{\psi}^m_{2i}\right) + 2J_7^m\left(\dot{\psi}^m_{3i}\right)\left(\dot{\psi}^f_{3i} - \dot{\psi}^m_{3i}\right)\Big\} , \qquad (2.37)$$

where A_c is the cross-sectional area of a cell

$$A_c = d_2 d_3 \qquad (2.38)$$

The strain energy density follows from (2.36) as

$$A_c W = \frac{1}{2}(\lambda_f + 2\mu_f)\left\{A_f(\partial_1 \bar{u}_1)^2 + I_3^f\left(\partial_1\psi^f_{21}\right)^2 + I_2^f\left(\partial_1\psi^f_{31}\right)^2 + A_f\left(\psi^f_{22}\right)^2 + A_f\left(\psi^f_{33}\right)^2\right\}$$

$$+ \lambda_f A_f\left\{(\partial_1\bar{u}_1)\left(\psi^f_{22}\right) + (\partial_1\bar{u}_1)\left(\psi^f_{33}\right) + \psi^f_{22}\,\psi^f_{33}\right\} + \frac{1}{2}\mu_f\left\{A_f\left(\partial_1\bar{u}_2 + \psi^f_{21}\right)^2\right.$$

$$+ I_3^f\left(\partial_1\psi^f_{22}\right)^2 + I_2^f\left(\partial_1\psi^f_{32}\right)^2 + A_f\left(\partial_1\bar{u}_3 + \psi^f_{31}\right)^2 + I_3^f\left(\partial_1\psi^f_{23}\right)^2 + I_2^f\left(\partial_1\psi^f_{33}\right)^2$$

$$\left. + A_f\left(\psi^f_{23} + \psi^f_{32}\right)^2\right\} + \frac{1}{2}(\lambda_m + 2\mu_m)\left\{A_m(\partial_1\bar{u}_1)^2 + I_3^m\left(\partial_1\psi^m_{21}\right)^2 + I_2^m\left(\partial_1\psi^m_{31}\right)^2\right.$$

$$+ J_1^m\left(\partial_1\psi^f_{21} - \partial_1\psi^m_{21}\right)^2 + J_2^m\left(\partial_1\psi^f_{31} - \partial_1\psi^m_{31}\right)^2 + 2J_6^m\left(\partial_1\psi^m_{21}\right)\left(\partial_1\psi^f_{21} - \partial_1\psi^m_{21}\right)$$

$$+ 2J_7^m\left(\partial_1\psi^m_{31}\right)\left(\partial_1\psi^f_{31} - \partial_1\psi^m_{31}\right) + A_m\left(\psi^m_{22}\right)^2 + J_4^m\left(\psi^f_{22} - \psi^m_{22}\right)^2 + J_5^m\left(\psi^f_{33} - \psi^m_{32}\right)^2$$

$$+ 2J_8^m\left(\psi^m_{22}\right)\left(\psi^f_{22} - \psi^m_{22}\right) + A_m\left(\psi^m_{33}\right)^2 + J_5^m\left(\psi^f_{23} - \psi^m_{23}\right)^2 + J_3^m\left(\psi^f_{33} - \psi^m_{33}\right)^2$$

$$\left. + 2J_9^m\left(\psi^m_{33}\right)\left(\psi^f_{33} - \psi^m_{33}\right)\right\} + \lambda_m\left\{A_m(\partial_1\bar{u}_1)\left(\psi^m_{22}\right) + J_8^m(\partial_1\bar{u}_1)\left(\psi^f_{22} - \psi^m_{22}\right)\right.$$

$$+ A_m\left(\partial_1 \bar{u}_1\right)\left(\psi^m_{33}\right) + J^m_9\left(\partial_1 \bar{u}_1\right)\left(\psi^f_{33} - \psi^m_{33}\right) + A_m\left(\psi^m_{22}\right)\left(\psi^m_{33}\right) + J^m_8\left(\psi^m_{33}\right)\left(\psi^f_{22} - \psi^m_{22}\right)$$

$$+ J^m_9\left(\psi^m_{22}\right)\left(\psi^f_{33} - \psi^m_{33}\right) + J^m_5\left(\psi^f_{22} - \psi^m_{22}\right)\left(\psi^f_{33} - \psi^m_{33}\right) + J^m_5\left(\psi^f_{32} - \psi^m_{32}\right)\left(\psi^f_{23} - \psi^m_{23}\right)\Big)\Big\}$$

$$+ \frac{1}{2}\mu_m\Big\{A_m\left(\partial_1 \bar{u}_2\right)^2 + A_m\left(\psi^m_{21}\right)^2 + J^m_1\left(\partial_1 \psi^f_{22} - \partial_1 \psi^m_{22}\right)^2 + J^m_2\left(\partial_1 \psi^f_{32} - \partial_1 \psi^m_{32}\right)^2$$

$$+ I^m_3\left(\partial_1 \psi^m_{22}\right)^2 + I^m_2\left(\partial_1 \psi^m_{32}\right)^2 + J^m_4\left(\psi^f_{21} - \psi^m_{21}\right)^2 + J^m_5\left(\psi^f_{31} - \psi^m_{31}\right)^2 + 2A_m\left(\partial_1 \bar{u}_2\right)$$

$$\times \left(\psi^m_{21}\right) + 2J^m_8\left(\partial_1 \bar{u}_2\right)\left(\psi^f_{21} - \psi^m_{21}\right) + 2J^m_8\left(\psi^m_{21}\right)\left(\psi^f_{21} - \psi^m_{21}\right) + 2J^m_6\left(\partial_1 \psi^m_{22}\right)$$

$$\times \left(\partial_1 \psi^f_{22} - \partial_1 \psi^m_{22}\right) + 2J^m_7\left(\partial_1 \psi^m_{32}\right)\left(\partial_1 \psi^f_{32} - \partial_1 \psi^m_{32}\right) + A_m\left(\partial_1 \bar{u}_3\right)^2 + A_m\left(\psi^m_{31}\right)^2$$

$$+ J^m_1\left(\partial_1 \psi^f_{23} - \partial_1 \psi^m_{23}\right)^2 + J^m_2\left(\partial_1 \psi^f_{33} - \partial_1 \psi^m_{33}\right)^2 + I^m_3\left(\partial_1 \psi^m_{23}\right)^2 + I^m_2\left(\partial_1 \psi^m_{33}\right)^2$$

$$+ J^m_5\left(\psi^f_{21} - \psi^m_{21}\right)^2 + J^m_3\left(\psi^f_{31} - \psi^m_{31}\right)^2 + 2A_m\left(\partial_1 \bar{u}_3\right)\left(\psi^m_{31}\right) + 2J^m_9\left(\partial_1 \bar{u}_3\right)\left(\psi^f_{31} - \psi^m_{31}\right)$$

$$+ 2J^m_9\left(\psi^m_{31}\right)\left(\psi^f_{31} - \psi^m_{31}\right) + 2J^m_6\left(\partial_1 \psi^m_{23}\right)\left(\partial_1 \psi^f_{23} - \partial_1 \psi^m_{23}\right) + 2J^m_7\left(\partial_1 \psi^m_{33}\right)$$

$$\times \left(\partial_1 \psi^f_{33} - \partial_1 \psi^m_{33}\right) + A_m\left(\psi^m_{23}\right)^2 + A_m\left(\psi^m_{32}\right)^2 + J^m_4\left(\psi^f_{23} - \psi^m_{23}\right)^2 + J^m_5\left(\psi^f_{33} - \psi^m_{33}\right)^2$$

$$+ J^m_5\left(\psi^f_{22} - \psi^m_{22}\right)^2 + J^m_3\left(\psi^f_{32} - \psi^m_{32}\right)^2 + 2A_m\left(\psi^m_{23}\right)\left(\psi^m_{32}\right) + 2J^m_8\left(\psi^m_{23}\right)\left(\psi^f_{23} - \psi^m_{23}\right)$$

$$+ 2J^m_9\left(\psi^m_{23}\right)\left(\psi^f_{32} - \psi^m_{32}\right) + 2J^m_8\left(\psi^m_{32}\right)\left(\psi^f_{23} - \psi^m_{23}\right) + 2J^m_9\left(\psi^m_{32}\right)\left(\psi^f_{32} - \psi^m_{32}\right)$$

$$+ 2J_5^m\left(\overset{f}{\psi}_{23} - \overset{m}{\psi}_{23}\right)\left(\overset{f}{\psi}_{32} - \overset{m}{\psi}_{32}\right) + 2J_5^m\left(\overset{f}{\psi}_{33} - \overset{m}{\psi}_{33}\right)\left(\overset{f}{\psi}_{22} - \overset{m}{\psi}_{22}\right) \quad , \tag{2.39}$$

where the constants are given by Eqs. (2.19), (2.20) and (2.24)-(2.34).

It remains to examine what happens to the continuity conditions (2.6) and (2.8) in the transition from the system of variables defined along discrete lines to the system of continuous variables. Considering $\bar{u}_i$, etc. as continuous functions of x_2 and x_3, Eq. (2.6) is a difference relation of the form

$$\Delta_k \bar{u}_i - \frac{\alpha\, d_3}{d_2}\, \ell n \left[\frac{1 + (1+\zeta^2)^{\frac{1}{2}}}{\zeta}\right]\left(2\, \overset{f}{\psi}_{3i} + \Delta_k \overset{f}{\psi}_{3i} - 2\, \overset{m}{\psi}_{3i} - \Delta_k \overset{m}{\psi}_{3i}\right) - \frac{1}{2} d_3 \left(2\, \overset{m}{\psi}_{3i} + \Delta_k \overset{m}{\psi}_{3i}\right) \tag{2.40}$$

In Eq. (2.40) the field variables are considered at $x_2 = \overset{\ell}{x}_2$, $x_3 = \overset{k}{x}_3$. The difference $\Delta_k \bar{u}_i$ is defined as

$$\Delta_k \bar{u}_i = \bar{u}_i \Big|_{x_3 = \overset{k+1}{x_3}} - \bar{u}_i \Big|_{x_3 = \overset{k}{x_3}} \quad , \tag{2.41}$$

with analogous definitions for $\Delta_k \overset{f}{\psi}_{3i}$ and $\Delta_k \overset{m}{\psi}_{3i}$. Noting that $\overset{k+1}{x_3} = \overset{k}{x_3} + d_3$, we see that in the limit $d_3 \to 0$, $d_2 \to 0$, but keeping $\zeta = d_3/d_2$ and α/d_2 constant, the difference relation (2.40) can be replaced by the differential relation

$$S_{3i} = \partial_3 \bar{u}_i - \frac{2\alpha}{d_2}\, \ell n \left[\frac{1 + (1+\zeta^2)^{\frac{1}{2}}}{\zeta}\right]\left(\overset{f}{\psi}_{3i} - \overset{m}{\psi}_{3i}\right) - \overset{m}{\psi}_{3i} = 0 \tag{2.42}$$

Similarly we obtain from Eq. (2.8)

$$S_{2i} = \partial_2 \bar{u}_i - \frac{2\alpha}{d_3}\, \ell n \left[\zeta + (1+\zeta^2)^{\frac{1}{2}}\right]\left(\overset{f}{\psi}_{2i} - \overset{m}{\psi}_{2i}\right) - \overset{m}{\psi}_{2i} = 0 \tag{2.43}$$

It is now assumed that (2.42) and (2.43) are also valid for finite values of d_2 and d_3. The continuity conditions in the system of discrete cells as given by Eqs. (2.6) and (2.8) have thus been turned into constraint conditions between the continuous field variables.

At this stage we have constructed a strain energy density as an expression in terms of local deformations and the gradients of the local deformations and the gross displacements. A kinetic energy density has been obtained in terms of the first order time derivatives of the gross displacements and the local deformations. Considering a fixed regular region V of the medium, the displacement equations of motion can then be obtained by invoking Hamilton's principle for independent variations of the dependent field quantities in V and in a specified time interval $t_0 \leq t \leq t_1$. For the region V, Hamilton's principle states that

$$\delta \int_{t_0}^{t_1}\int_{V} (T-W)\,dt\,dV + \int_{t_0}^{t_1} \delta W_1\,dt = 0\ , \tag{2.44}$$

where δW_1 is the variation of the work done by external forces and dV is the scalar volume element. Here we are interested only in the displacement equations of motion and we restrict the admissible variations to ones that vanish identically on the bounding surface of V. In the absence of body forces the variational problem then reduces to finding the Euler equations for

$$\delta \int_{t_0}^{t}\int_{V} F\,dt\,dV = 0\ , \tag{2.45}$$

where the functional F is defined as

$$F = T - W \tag{2.46}$$

An elegant and convenient method of taking the continuity conditions (2.42) and (2.43) into account is to introduce them as subsidiary conditions through the use of Lagrangian multipliers. The variational problem may then be redefined by using the functional

$$F = T - W - \sum_{i=1}^{3}\left(\Gamma_{2i}\,S_{2i} + \Gamma_{3i}\,S_{3i}\right)\ , \tag{2.47}$$

in Eq. (2.45), where the Lagrangian multipliers Γ_{2i} and Γ_{3i} are functions of x_j and t, and S_{2i} and S_{3i} are defined by Eqs (2.42) and (2.43), respectively. Since the functional F as given by Eq. (2.47) depends only on the dependent field variables and their first order derivatives the system of Euler equations may be written as

$$\sum_{r=1}^{4}\frac{\partial}{\partial q_r}\left[\frac{\partial F}{\partial(\partial f_s/\partial q_r)}\right] - \frac{\partial F}{\partial f_s} = 0 \tag{2.48}$$

In Eq (2.48), f_s represents the twelve dependent variables $\bar{u}_i$, ψ_{2i}^{f}, ψ_{2i}^{m} and Γ_i , and q_r are the spatial variables x_i and time t. A system of 21 governing equations follows from the Euler equation (2.48) and from the constraint conditions (2.42) and (2.43).

2.3. Transverse Waves Propagating Parallel to the Fibers

Let us consider motions which depend on x_1 and t only, and which are described by

$$\bar{u}_2(x_1,t),\ \overset{f}{\psi}_{21}(x_1,t),\ \overset{m}{\psi}_{21}(x_1,t) \text{ and } \Gamma_{21}(x_1,t) \tag{2.49}$$

These field variables describe transverse motions in the x_2-direction. It can be verified from the expressions for the kinetic and strain energy densities, Eqs. (2.37) and (2.39), respectively, that the fields described by (2.49) are not coupled to other fields. Thus motions described by (2.49) are indeed possible according to the present theory. For fields of the form (2.49) the energy densities simplify considerably. Application of Eq. (2.48) yields four relatively simple equations for the four field variables of Eq. (2.49). The Equations are

$$a_1\partial_1\partial_1\bar{u}_2 + a_2\partial_1\overset{f}{\psi}_{21} + a_3\partial_1\overset{m}{\psi}_{21} = a_4\ddot{\bar{u}}_2 \tag{2.50}$$

$$a_5\partial_1\bar{u}_2 + a_6\partial_1\partial_1\overset{f}{\psi}_{21} + a_7\partial_1\partial_1\overset{m}{\psi}_{21} + a_8\overset{f}{\psi}_{21} + a_9\overset{m}{\psi}_{21} + a_{10}\Gamma_{21} = a_{11}\ddot{\overset{f}{\psi}}_{21} + a_{12}\ddot{\overset{m}{\psi}}_{21} \tag{2.51}$$

$$a_{13}\partial_1\bar{u}_2 + a_{14}\partial_1\partial_1\overset{f}{\psi}_{21} + a_{15}\partial_1\partial_1\overset{m}{\psi}_{21} + a_{16}\overset{f}{\psi}_{21} + a_{17}\overset{m}{\psi}_{21} + a_{18}\Gamma_{21} = a_{19}\ddot{\overset{f}{\psi}}_{21} + a_{20}\ddot{\overset{m}{\psi}}_{21} \tag{2.52}$$

$$a_{21}\overset{f}{\psi}_{21} + a_{22}\overset{m}{\psi}_{21} = 0 \tag{2.53}$$

The constants a_1 ----- a_{22} are

$$a_1 = \mu_f A_f + \mu_m A_m$$

$$a_2 = \mu_f A_f + \mu_m \overset{m}{J}_8$$

$$a_3 = \mu_m A_m - \mu_m \overset{m}{J}_8$$

$$a_4 = \rho_f A_f + \rho_m A_m$$

$$a_5 = -\mu_f A_f - \mu_m J_8^m$$

$$a_6 = (\lambda_f + 2\mu_f) I_3^f + (\lambda_m + 2\mu_m) J_1^m$$

$$a_7 = \left(\lambda_m + 2\mu_m\right)\left(J_6^m - J_1^m\right)$$

$$a_8 = -\left(\mu_f A_f + \mu_m J_4^m + \mu_m J_5^m\right)$$

$$a_9 = \mu_m J_4^m - \mu_m J_8^m + \mu_m J_5^m$$

$$a_{10} = 2\alpha d_2 \cdot \ln\left[\zeta + (1+\zeta^2)^{\frac{1}{2}}\right]$$

$$a_{11} = \rho_f I_3^f + \rho_m J_1^m$$

$$a_{12} = \rho_m J_6^m - \rho_m J_1^m$$

$$a_{13} = -\mu_m\left(A_m - J_8^m\right)$$

$$a_{14} = \left(\lambda_m + 2\mu_m\right)\left(J_6^m - J_1^m\right)$$

$$a_{15} = \left(\lambda_m + 2\mu_m\right)\left(I_3^m + J_1^m - 2J_6^m\right)$$

$$a_{16} = \mu_m\left(J_4^m - J_8^m + J_5^m\right)$$

$$a_{17} = \mu_m\left(-A_m - J_4^m + 2J_8^m - J_5^m\right)$$

$$a_{18} = A_c - 2\alpha d_2 \ln\left[\zeta + (1+\zeta^2)^{\frac{1}{2}}\right]$$

$$a_{19} = \rho_m\left(J_6^m - J_1^m\right)$$

$$a_{20} = \rho_m\left(J_1^m + I_3^m - 2J_6^m\right)$$

$$a_{21} = -2(\alpha/d_3)\ln\left[\zeta + (1+\zeta^2)^{\frac{1}{2}}\right]$$

$$a_{22} = -1 + 2(\alpha/d_3)\ln\left[\zeta + (1+\zeta^2)^{\frac{1}{2}}\right]$$

The system of displacement equations of motion (2.50)-(2.53) may be used to study the propagation of transverse waves propagating in the direction of the fibers. In the context of the present theory such waves are defined by

$$\bar{u}_2 = B_1 \exp[\, ik\,(x_1 - ct)] \tag{2.54}$$

$$\psi^{f}_{21} = B_2 \exp[\, ik\,(x_1 - ct)] \tag{2.55}$$

$$\psi^{m}_{21} = B_3 \exp[\, ik\,(x_1 - ct)] \tag{2.56}$$

$$\Gamma_{21} = B_4 \exp[\, ik\,(x_1 - ct)] \tag{2.57}$$

In Eqs. (2.54)-(2.57), B_1—B_4 are constant amplitudes, k is the wavenumber (wavenumber = 2π/wavelength), and c is the phase velocity. Substitution of (2.54)-(2.57) into the equations of motion (2.50)-(2.53) yields four homogeneous equations for B_1, B_2, B_3 and B_4. For a nontrivial set of solutions the determinant of the coefficients must vanish. This yields a relation between the phase velocity c and wavenumber k, which is called the dispersion relation.

Numerical results for the phase velocity as a function of the wavenumber were worked out for three values of μ_f/μ_m, namely

$$\frac{\mu_f}{\mu_m} = 100\ , \quad \frac{\mu_f}{\mu_m} = 50\ , \quad \frac{\mu_f}{\mu_m} = 10$$

The other material parameters were chosen as

$$\frac{\rho_f}{\rho_m} = 3\ , \quad \nu_f = 0.3\ , \quad \nu_m = 0.35\ ,$$

where ν_f and ν_m denote Poisson's ratio in the materials of the fibers and the matrix, respectively. The parameters defining the geometrical layout were chosen as

$$\frac{d_3}{d_2} = 1 \text{ and } \frac{\alpha}{d_2} = 0.4$$

Curves for the dimensionless phase velocity versus the dimensionless wavenumber are shown in Fig. 2.6. It is noted that the curves reveal a very marked dispersive behavior, even at small values of $k\alpha$, and especially for large values of μ_f/μ_m. Note that for $\mu_f/\mu_m = 100$ and at $k\alpha = 1$, which corrsponds to a wavelength of π times the diameter of a fiber, the phase velocity is almost three times the phase velocity at $k\alpha = 0$.

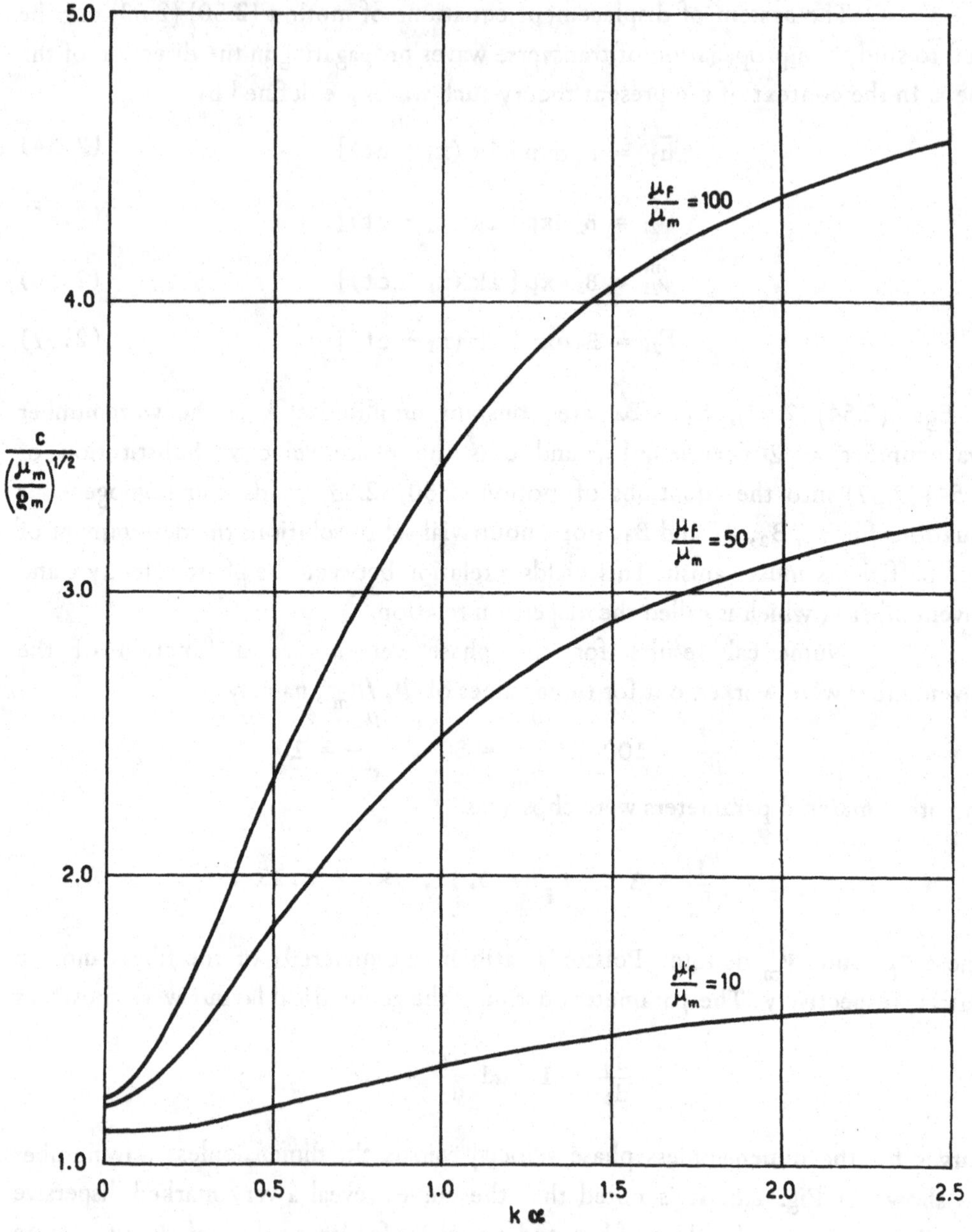

Fig. 2.6. Dimensionless phase velocity versus dimensionless wavenumber for transverse waves ; $d_2/d_3 = 1$, $\alpha/d_2 = 0.4$, $\rho_f/\rho_m = 3$, $\nu_f = 0.3$, $\nu_m = 0.35$

REFERENCES

[2.1] C.-T. Sun, J.D. Achenbach and G. Herrmann, J. Appl. Mech. 35, p. 467 (1968)

[2.2] J.D. Achenbach, C.-T. Sun and G. Herrmann, J. Appl. Mech, 35 p. 689 (1968).

[2.3] E. and F. Cosserat, Théorie des Corps Deformables, A. Hermann et Fils, Paris (1909).

[2.4] R.D. Mindlin and H.F. Tiersten, Arch.Rat. Mech. Anal. 11, p. 415 (1962).

[2.5] R. Stojanović, Mechanics of Polar Continua CISM, Udine (1969)

[2.6] R. Stojanović, Recent Developments in the Theory of Polar Continua, CISM, Udine (in press).

[2.7] G. Herrmann and J.D. Achenbach, in Mechanics of Generalized Continua (E. Kröner, ed.), Springer Verlag, Berlin (1968).

[2.8] R.D. Mindlin, Arch. Rat. Mech. Anal. 16, p. 51 (1964).

[2.9] R.A. Grot and J.D. Achenbach, Acta Mechanica 9, p. 245 (1970).

[2.10] R.A. Grot and J.D. Achenbach, Int. J. Solids and Structures 6, p. 641 (1970).

[2.11] J.D. Achenbach and C.-T. Sun, in Dynamics of Composite Materials, (E.H. Lee, ed), American Society of Mechanical Engineers, New York (1972).

PART III

ADDENDUM
SPECIAL THEORIES AND COMPARISONS WITH EXPERIMENTAL RESULTS

PREFACE TO THE ADDENDUM

Parts I and II of this monograph were completed in October 1973. When the publication of the work was delayed until the end of 1975, it was thought desirable to update the manuscript by adding some examples of the theory in this Addendum. Some special cases are discussed, and the analytical results are compared with ultrasonic test data, and with some results obtained by the finite element technique.

Evanston, January 1976.

CHAPTER 1
THE LAMINATED MEDIUM

1.1 Introduction

An important motivation for a detailed study of the propagation of harmonic waves in laminated media and fiber reinforced composites is that effective elastic constants can conveniently be measured by ultrasonic testing techniques. These techniques have the advantage that small specimens can be used, and that good control and reproducibility can be achieved. Typically one measures phase velocity or group velocity for a number of frequencies. The extrapolation to zero frequency then provides the velocity for very long waves, from which an effective elastic constant can be determined.

The interpretation of ultrasonic test results requires a good understanding of the dynamic behavior of composite materials. The last part of this monograph is, therefore, concerned with wave motions that are relevant to ultrasonic testing techniques. A number of specific examples involving the dispersive behavior of time-harmonic waves are discussed. The results are compared with some recent experimental results. Chapter 1 is concerned with laminated media. Fiber-reinforced composites are discussed in Chapter 2.

The propagation of harmonic waves in a layered composite consisting of alternating layers of two elastic materials can be analyzed rigorously, see e.g. Refs. 3.1, 3.2 and 3.3. It is also quite simple to construct an effective modulus theory for a laminated medium, as shown in Ref. 3.4. Thus, a laminated composite provides a very suitable model to display the limitations of the effective modulus theory. This was done in Ref. 3.5.

The exact results for a laminated medium presented in Ref. 3.3 exhibit the different nature of the dispersive behavior for harmonic waves propagating in the direction of the layering, and normal to the layering. For waves propagating along the layering, the layers act as waveguides, and there are no stop-bands, i.e., frequency ranges in which propagating harmonic waves are not possible. For waves propagating normal to the layering, the dynamic interaction between neighboring layers does generate stop-bands, which are very similar to those found in elastic lattices (see e.g. Ref. 3.6).

Equations governing a homogeneous continuum model for a laminated medium were presented in Part II. Section 1.4 of Part II also contains comparisons of the dispersion curves with exact results. The comparisons show that for transverse waves a model based on two-term expansions of the displacements yields good results for waves propagating in the direction of the layering. In Ref. 3.5 it was shown that for waves propagating normal to the direction of the layering, a two-term model yields good results only for relatively small values of the ratio of the characteristic length of the structuring (thickness of a pair of layers) over the characteristic length of the deformation (the wavelength).

In this Chapter a more accurate homogeneous continuum model for a laminated composite is constructed, by including quadratic terms in the displacement expansions, and by representing the interaction between neighboring cells in a more accurate manner. We will present the details for one-dimensional deformations, with displacements normal to the layering and parallel to the layering, respectively. For transverse waves propagating normal to the layering, analytical results showing the relation between phase velocity and wavenumber are compared with experimental results from Ref. 3.7

1.2 <u>One-Dimensional Deformations Propagating Normal to the Layering</u>

For both purely longitudinal and purely transverse wave motions propagating normal to the layering, the field variables depend only on one spatial coordinate x, and on time. The one non-vanishing displacement component for these cases is denoted by u(x,t). For longitudinal motions u(x,t) is parallel to x; for transverse motions u(x,t) is perpendicular to x. The geometry for the system of alternate layers of solids 1 and 2 is shown in Fig. 3.1. Note that the layer of material 1 is taken as the core of cell k.

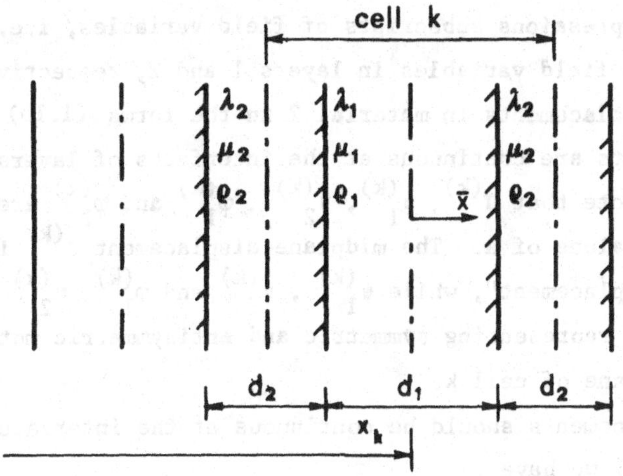

Fig. 3.1 Cell k of laminated composite.

To construct an appropriate homogeneous continuum model, the displacement in cell k is approximated by expansions over the local coordinate $\bar{x}$, in terms of quantities which are defined in the midplane of the cell. We write

$|\bar{x}| \le \frac{1}{2} d_1$:

$$u_1^{(k)} = \bar{u}^{(k)}(x^k,t) + \bar{x}\psi_1^{(k)}(x^k,t) + \frac{\bar{x}^2}{d_1}\varphi_1^{(k)}(x^k,t) \tag{1.1a}$$

$\frac{1}{2} d_1 \le \bar{x} \le \frac{1}{2}(d_1+d_2)$:

$$u_2^{(k)} = \bar{u}^{(k)}(x^k,t) + \frac{1}{2} d_1\psi_1^{(k)}(x^k,t) + (\bar{x} - \frac{1}{2}d_1)\psi_2^{(k)}(x^k,t)$$

$$+ \frac{1}{4}d_1\,\varphi_1^{(k)}(x^k,t) + \frac{\left(\bar{x}-\frac{1}{2}d_1\right)^2}{d_2}\varphi_2^{(k)}(x^k,t) \tag{1.1b}$$

$-\frac{1}{2}(d_1+d_2) \le \bar{x} \le -\frac{1}{2}d_1$:

$$u_2^{(k)} = \bar{u}^{(k)}(x^k,t) - \frac{1}{2}d_1\psi_1^{(k)}(x^k,t) + (\bar{x} + \frac{1}{2}d_1)\psi_2^{(k)}(x^k,t)$$

$$+ \frac{1}{4}d_1\ \varphi_1^{(k)}(x^k,t) + \frac{\left(\bar{x} + \frac{1}{2}d_1\right)^2}{d_2}\varphi_2^{(k)}(x^k,t) \tag{1.1c}$$

Note that in these expressions subscripts of field variables, i.e., 1 and 2, define these field variables in layers 1 and 2, respectively. By expressing the displacements in material 2 in the forms (1.1b) and (1.1c), the displacements are continuous at the interfaces of layers of materials 1 and 2. Note that $\bar{u}^{(k)}$, $\psi_1^{(k)}$, $\psi_2^{(k)}$, $\varphi_1^{(k)}$ and $\varphi_2^{(k)}$ are defined at discrete values of x. The midplane displacement $\bar{u}^{(k)}$ is called the "gross displacement", while $\psi_1^{(k)}$, $\psi_2^{(k)}$ and $\varphi_1^{(k)}$, $\varphi_2^{(k)}$ are "local deformations", representing symmetric and antisymmetric motions relative to the midplane of cell k.

Since the displacements should be continuous at the interface of cell k and cell k+1 we have

$$u_2^{(k+1)}\Big|_{\bar{x} = -\frac{1}{2}(d_1+d_2)} = u_2^{(k)}\Big|_{\bar{x} = \frac{1}{2}(d_1+d_2)} \tag{1.2}$$

By substituting Eqs. (1.1a-c) in Eq. (1.2) we obtain

$$\Delta\bar{u}^{(k)} - d_1\psi_1^{(k)} - \frac{1}{2}d_1\Delta\psi_1^{(k)} - d_2\psi_2^{(k)} - \frac{1}{2}d_2\Delta\psi_2^{(k)}$$

$$+ \frac{1}{4}d_1\Delta\varphi_1^{(k)} + \frac{1}{4}d_2\Delta\varphi_2^{(k)} = 0 \tag{1.3}$$

Here the difference symbol Δ is defined as

$$\Delta\bar{u}^{(k)} = \bar{u}^{(k+1)} - \bar{u}^{(k)}\ , \text{ etc.} \tag{1.4}$$

The strain components $\epsilon_1^{(k)}$ and $\epsilon_2^{(k)}$ are computed from Eqs. (1.1a-c) by differentiation with respect to $\bar{x}$. The strain energy in the reinforcing element of cell k can then be written as:

for longitudinal motions:

$$W_1^{(k)} = \frac{1}{2}\left(\lambda_1+2\mu_1\right)d_1\left(\psi_1^{(k)}\right)^2 + \frac{1}{6}\left(\lambda_1+2\mu_1\right)d_1\left(\varphi_1^{(k)}\right)^2$$

for transverse motions:

$$W_1^{(k)} = \frac{1}{2}\mu_1 d_1\left(\psi_1^{(k)}\right)^2 + \frac{1}{6}\mu_1 d_1\left(\varphi_1^{(k)}\right)^2$$

Where λ_1 and μ_1 are the Lamé elastic constants for the material of layer 1. Analogous expressions can be found for the strain energy stored in material 2 of cell k. These expressions are

$$W_2^{(k)} = \frac{1}{2}\left(\lambda_2+2\mu_2\right) d_2\left(\psi_2^{(k)}\right)^2 + \frac{1}{6}\left(\lambda_2+2\mu_2\right) d_2\left(\varphi_2^{(k)}\right)^2$$

and

$$W_2^{(k)} = \frac{1}{2}\mu_2 d_2\left(\psi_2^{(k)}\right)^2 + \frac{1}{6}\mu_2 d_2\left(\varphi_2^{(k)}\right)^2$$

The average over the volume of cell k is

$$W_{ave}^{(k)} = \frac{W_1^{(k)} + W_2^{(k)}}{d_1 + d_2} \tag{1.5}$$

In computing the kinetic energy stored in cell k we only take the gross displacement $u^{(k)}$ into account. The average over the volume of cell k is easily obtained as

$$T = \frac{1}{2}\bar{\rho}\left(\dot{u}^{(k)}\right)^2, \tag{1.6}$$

where

$$\bar{\rho} = \eta\rho_1 + (1-\eta)\rho_2 \tag{1.7}$$

$$\eta = \frac{d_1}{d_1+d_2} \tag{1.8}$$

The one-dimensional motion of the laminated composite is now described by the field variables $u^{(k)}$, $\psi_1^{(k)}$, $\psi_2^{(k)}$, $\varphi_1^{(k)}$ and $\varphi_2^{(k)}$, which are defined only in discrete parallel planes $x = x^k$. To construct a homogeneous continuum model for the laminated medium, we introduce fields $u(x,t)$, $\psi_1(x,t)$, $\psi_2(x,t)$, $\varphi_1(x,t)$ and $\varphi_2(x,t)$, and we effect the transition to the continuum model as discussed in Part II. For the strain energy density we find

$$W = \frac{1}{2}a_1(\psi_1)^2 + \frac{1}{2}a_2(\psi_2)^2 + \frac{1}{2}b_1(\varphi_1)^2 + \frac{1}{2}b_2(\varphi_2)^2 \tag{1.9}$$

where for longitudinal motions

$$a_1 = (\lambda_1+2\mu_1)\eta \quad ; \quad a_2 = (\lambda_2+2\mu_2)(1-\eta)$$

and for transverse motions

$$a_1 = \mu_1\eta \quad ; \quad a_2 = \mu_2(1-\eta)$$

In both cases

$$b_1 = \frac{1}{3}a_1 \qquad b_2 = \frac{1}{3}a_2$$

The kinetic energy density is

(1.10) $$T = \frac{1}{2}\bar{\rho}\,\dot{\bar{u}}^2$$

where $\bar{\rho}$ is defined by Eq. (1.7)

To place Eq. (1.3) within the context of the transition to the continuum model, we introduce Taylor expansions for the differences $\Delta\bar{u}^{(k)} = \bar{u}^{(k+1)} - \bar{u}^{(k)}$, etc. Defining the operator P[ ] as

(1.11) $$P[\ \ ] = \sum_{n=1}^{\infty} \frac{1}{n!}\left(d_1+d_2\right)^n \frac{\partial^n}{\partial x^n}$$

we find that Eq. (1.3) can be replaced by

(1.12) $$S = P[\bar{u}] - d_1\psi_1 - \frac{1}{2}d_1P[\psi_1] - d_2\psi_2 - \frac{1}{2}d_2P[\psi_2] + \frac{1}{4}d_1P[\varphi_1] + \frac{1}{4}d_2P[\varphi_2] = 0$$

The functional to be used in Hamilton's principle is now defined as

(1.13) $$F = T - W - \gamma S$$

where γ is a Lagrangian multiplier.

To obtain the Euler-Poisson equation corresponding to Eq. (1.23) of Part II, with F defined by Eq. (1.13), we employ a well-known result which states that the Euler-Poisson equation for

$$I = \int F(x;y_1,y_1^{(1)},\ldots y_1^{(n_1)};\ y_2\ldots y_2^{(n_2)};\ldots;y_m\ldots y_m^{(n_m)})dx$$

is given by

(1.14) $$\sum_{k=0}^{n_i}\left(-1\right)^k \frac{d^k}{dx^k}\left(\frac{\partial F}{\partial y_i^{(k)}}\right) = 0$$

Here $y_1\ldots y_m$ are functions of x, and $y_i^{(k)}$ denotes d^ky_i/dx^k. Application of Eq. (1.14) to Eq. (1.13), where T, W and S are defined by Eqs. (1.10), (1.9), and (1.12), respectively, yields

(1.15) $$-\bar{\rho}\,\ddot{\bar{u}} - Q[\gamma] = 0$$

(1.16) $$-a_1\psi_1 + d_1\gamma + \frac{1}{2}d_1\,Q[\gamma] = 0$$

$$-a_2\psi_2 + d_2\gamma + \frac{1}{2}d_2\, Q[\gamma] = 0 \tag{1.17}$$

$$-b_1\varphi_1 - \frac{1}{4}d_1\, Q[\gamma] = 0 \tag{1.18}$$

$$-b_2\varphi_2 - \frac{1}{4}d_2\, Q[\gamma] = 0 \tag{1.19}$$

$$S = 0 \tag{1.20}$$

In these equations the operator Q[ ] is defined as

$$Q[\;] = \sum_{n=1}^{\infty} \frac{1}{n!}(d_1+d_2)^n(-1)^n \frac{\partial^n}{\partial x^n} \tag{1.21}$$

Now let us consider harmonic waves propagating in the x-direction. For these waves the field variables are of the forms

$$(u,\gamma) = (U,\Gamma)e^{ikx-i\omega t}$$

$$(\psi_1,\psi_2) = (\Psi_1,\Psi_2)e^{ikx-i\omega t} \tag{1.22}$$

$$(\varphi_1,\varphi_2) = (\Phi_1,\Phi_2)e^{ikx-i\omega t}$$

where $k = 2\pi/\Lambda$ is the wavenumber (Λ = wavelength), and ω is the circular frequency. Substitution of Eqs. (1.22) into Eqs. (1.15)-(1.20) yields a system of homogeneous algebraic equations for the constants U, Γ, Ψ_1, Ψ_2, Φ_1 and Φ_2. In these equations the operation $Q[\gamma]$ yields

$$Q[\Gamma\, e^{ikx-i\omega t}] = \Gamma\, e^{ikx-i\omega t} \sum_{n=1}^{\infty} \frac{1}{n!}(d_1+d_2)^n\,(-ik)^n$$

$$= \Gamma\, e^{ikx-i\omega t}(e^{-i\xi} - 1) \tag{1.23}$$

The operator P[ ], which appears in Eq. (1.20), yields

$$P[u] = P[Ue^{ikx-i\omega t}] = U\, e^{ikx-i\omega t}(e^{i\xi}-1) \tag{1.24}$$

In Eqs. (1.23) and (1.24) the dimensionless wavenumber ξ is defined as

$$\xi = k(d_1+d_2) \tag{1.25}$$

The condition that the determinant of the coefficients of the system of equations for U, Γ, Ψ_1, Ψ_2, Φ_1 and Φ_2 must vanish, now yields a very simple relation between the dimensionless frequency and the dimensionless wavenumber:

$$\Omega = \left[\frac{16(1-\cos\xi)}{4(1+\cos\xi) + 3(1-\cos\xi)}\right]^{\frac{1}{2}} \tag{1.26}$$

where

$$\Omega = \left(\frac{\rho}{C}\right)^{\frac{1}{2}} (d_1+d_2)\omega \tag{1.27}$$

For longitudinal motions the constant C in Eq. (1.27) is

$$C = C^*_{22} = \frac{(\lambda_1+2\mu_1)(\lambda_2+2\mu_2)}{\eta(\lambda_2+2\mu_2) + (1-\eta)(\lambda_1+2\mu_1)} \tag{1.28}$$

For transverse motions we have

$$C = C^*_{44} = \frac{\mu_1\mu_2}{\eta\mu_2 + (1-\eta)\mu_1} \tag{1.29}$$

These constants are recognized as the effective elastic moduli for one-dimensional deformations normal and parallel to the layering, respectively, see Eqs. (3.40) and (3.44) of Part I.

Equation (1.26) shows that for small values of k the frequency increases linearly with k, the proportionality factor being $(C/\bar{\rho})^{\frac{1}{2}}$. As the wavenumber increases the curve relating Ω and ξ becomes concave. A maximum value of Ω is reached when $d\Omega/d\xi = 0$, i.e., when

$$\frac{\sin\xi}{(1-\cos\xi)^{\frac{1}{2}}} = 0 \quad \text{or} \quad k = \frac{\pi}{d_f+d_m}. \tag{1.30}$$

The exact dispersion relation for one-dimensional wave motions normal to the layering has been given in Refs. 3.1 and 3.2. It is also included in Ref. 3.3. The frequency ω and the wavenumber k are related through the equation.

$$\cos\xi = \cos\frac{\omega d_1}{c_1}\cos\frac{\omega d_2}{c_2} - \frac{1}{2}\left(\zeta + \frac{1}{\zeta}\right)\sin\frac{\omega d_1}{c_1}\sin\frac{\omega d_2}{c_2} \tag{1.31}$$

For longitudinal waves we have

$$c_1 = \left(\frac{\lambda_1+2\mu_1}{\rho_1}\right)^{\frac{1}{2}} \; ; \quad c_2 = \left(\frac{\lambda_2+2\mu_2}{\rho_2}\right)^{\frac{1}{2}} \tag{1.32}$$

and for transverse waves

$$c_1 = \left(\frac{\mu_1}{\rho_1}\right)^{\frac{1}{2}} \; ; \quad c_2 = \left(\frac{\mu_2}{\rho_2}\right)^{\frac{1}{2}} \tag{1.33}$$

The impedance ratio ζ is defined as

$$\zeta = \frac{\rho_1 c_1}{\rho_2 c_2} \tag{1.34}$$

Typical solutions of Eq. (1.31) are shown in Fig. 1.7. In an homogeneous material, the relation between frequency and wavenumber would be a straight line. The most important feature of the solutions of (1.31) are the discontinuities or band gaps near $k = \pi/(d_1+d_2)$, $2\pi/(d_1+d_2)$, etc. At frequencies in these gaps, the number k becomes complex,and propagating waves are sharply attenuated in space, so the composite becomes effectively opaque to such signals. It should be noted that the approximate solution given by Eq. (1.26) describes this effect for the lowest band, see Eq. (1.30).

For longitudinal waves calculations were carried out for a laminated medium consisting of alternate layers of two materials for which the values of certain pertinent ratios are

$$\eta = \frac{d_1}{d_1+d_2} = 0.5 \quad ; \quad \mu_1/\mu_2 = 100$$

$$\rho_1/\rho_2 = 3. \quad ; \quad \nu_1 = \nu_2 = 0.3$$

Numerical values according to Eq. (1.26), and values computed from Eq. (1.31) are listed in Table 3.1.

Table 3.1: Approximate and Exact Frequencies for Longitudinal Waves

	$\Omega_L = (d_1+d_2)(\overline{\rho}/C^*_{22})^{\frac{1}{2}}\ \omega$	
$\xi = (d_1+d_2)k$	Eq. (1.26)	From Eq. (1.31)
0.628	0.62523	0.62257
1.257	1.23022	1.20868
1.885	1.76928	1.71162
2.513	2.16210	2.06598
2.827	2.27162	2.16325
3.142	2.30940	2.19661

The difference between approximate and exact frequencies does not exceed about 5%.

The relation between the frequency and the wavenumber given by Eq. (1.26) does have shortcomings. The most obvious one is that in the limit of either $\gamma = 1$, $\theta = 1$, or $\eta = 0$, or $\eta = 1$, equation (1.26) does not reduce to the linear relation between frequency and wavenumber which holds for homogeneous solids. The reason is that the interface condition given by Eq. (1.3) imposes a periodic structuring on the solid, which does not disappear in the limit of homogeneity of the material. It may, therefore, be expected that Eq. (1.26) will be less accurate, the more homogeneous the material is.

It is possible to improve the approximate relation between frequency and wavenumber by introducing a correction factor. For example, we can replace Eq. (1.12) by

$$S = P[u] - d_1\psi_1 - \frac{1}{2}d_1P[\psi_1] - d_2\,\psi_2 - \frac{1}{2}d_2\,P[\psi_2]$$

$$+ \varkappa\left\{\frac{1}{4}d_1\;P[\varphi_1] + \frac{1}{4}d_2\;P[\varphi_2]\right\} = 0 \tag{1.35}$$

where $\varkappa$ is the correction factor. Instead of Eq. (1.26) we now find

$$\Omega = \left[\frac{16(1-\cos\xi)}{4(1+\cos\xi) + 3\varkappa^2(1-\cos\xi)}\right]^{\frac{1}{2}} \tag{1.36}$$

The correction factor $\varkappa$ is computed from a comparison of Eq. (1.36) with either experimental results, or exact analytical results.

Experimental results for ultrasonic transverse waves propagating normal to the layering in a laminated steel/coppper composite have been presented in Ref. 3.7. The geometrical parameters for this composite are $d_1 = d_2 = 4.57 \times 10^{-3}$ cm. By using standard values for the material properties of steel and copper we obtain $\left(C_{44}^*/\bar{\rho}\right)^{\frac{1}{2}} = 2600$ m/sec. Substituting these values in Eq. (1.27) we find $\omega = 4.535\ \Omega$. The experimental values from Ref. 3.7 are listed in Table 3.2.

Table 3.2. Approximate Analytical and Experimental Frequencies for Transverse Waves

$\xi = (d_1+d_2)k$	ω (Megahertz) Experimental	Eq. (1.36), $\varkappa^2 = 0.614$
2.638	11.77	12.48
2.804	12.44	12.94
2.974	13.10	13.24
3.145	13.34	13.34

CHAPTER 2

THE FIBER-REINFORCED COMPOSITE

2.1 Introduction

Exact elastodynamic solutions within the context of classical elasticity theory are not available for fiber-reinforced composites. It is, however, to be expected that qualitatively a difference analogous to the one observed in laminated solids, should exist for dispersion of waves propagating in the direction of the fibers, and normal to the fiber-direction. This expectation has been confirmed by experimental results presented in Refs. 3.8 and 3.9.

The model described in Part II can represent the effects of the discrete structuring of a fiber-reinforced composite on the mechanical field variables. The model should be superior to the effective modulus theory, especially for dynamic problems.

One check on the accuracy of the model is a comparison with experimental results of the phase velocity at various frequencies, for specific harmonic wave motions. This comparison shows that linear expansions within a cell give good results, over a substantial range of frequencies, for transverse waves propagating in the direction of the fibers. The analytical results are valid over a smaller range of frequencies for longitudinal waves propagating in the direction of the fibers, and over a still smaller range for longitudinal waves propagating normal to the fibers.

In this chapter it is shown that the homogeneous continuum model can be improved by using more accurate displacement distributions, and by improving the representation of the interaction between neighboring cells.

A number of specific examples involving the dispersive behavior of time-harmonic waves propagating in directions parallel and normal to the fibers are discussed in this chapter. The results are compared with some recent experimental results, and with results from other theories, including some obtained by finite element techniques.

2.2 Transverse waves propagating in the direction of the fibers

For a number of frequencies, measurements of what is thought to be the group velocity, have been presented by Tauchert and Guzelsu, Ref. 3.8.

A simple theory for transverse waves can be based on the following assumed displacement representations in cell (k,ℓ), see Fig. 3.2: in fiber (k,ℓ), $(r \leq a)$:

(2.1) $$u_1^{f(k,\ell)} = \bar{x}_2\, \psi_{21}^{f(k,\ell)}$$

(2.2) $$u_2^{f(k,\ell)} = \bar{u}_2^{(k,\ell)}$$

in the matrix material of cell (k,ℓ):

(2.3) $$u_2^{m(k,\ell)} = \bar{u}_2^{(k,\ell)}$$

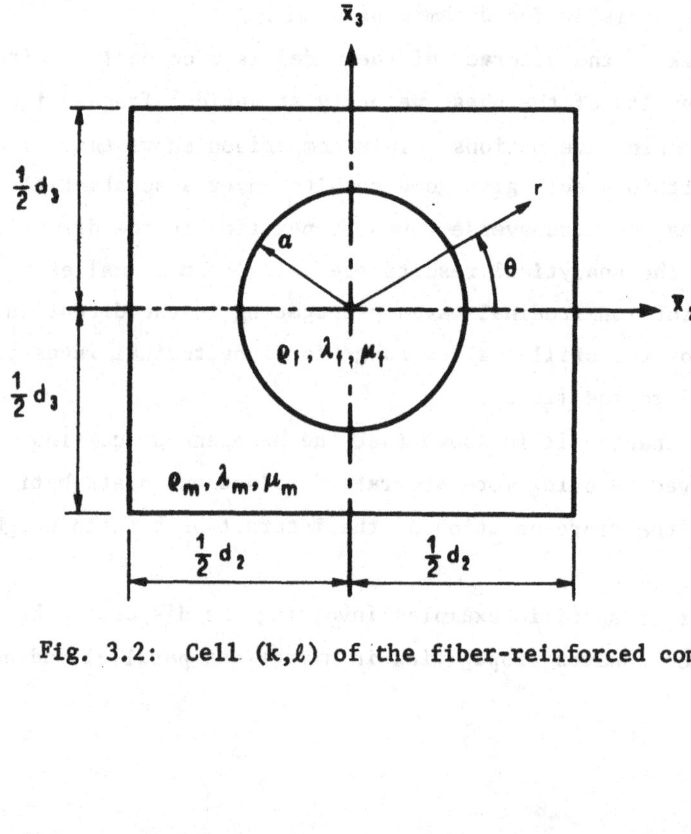

Fig. 3.2: Cell (k,ℓ) of the fiber-reinforced composite

$\bar{x}_2 \geq 0, |\bar{x}_3| \leq a$:

$$u_1^{m(k,\ell)} = (a^2-\bar{x}_3^2)^{\frac{1}{2}}\,\psi_{21}^{f(k,\ell)} + \left[\bar{x}_2 - \left(a^2 - \bar{x}_3^2\right)^{\frac{1}{2}}\right]\psi_{21}^{m(k,\ell)}$$

$a \leq |\bar{x}_3| \leq \frac{1}{2}d_3$:

$$u_1^{m(k,\ell)} = \bar{x}_2\psi_{21}^{m(k,\ell)}$$

$\bar{x}_2 \leq 0, |\bar{x}_3| \leq a$:

$$u_1^{m(k,\ell)} = -\left(a^2-\bar{x}_3^2\right)^{\frac{1}{2}}\psi_{21}^{m(k,\ell)} + \left[\bar{x}_2 + \left(a^2 - \bar{x}_3^2\right)^{\frac{1}{2}}\right]\psi_{21}^{m(k,\ell)}$$

(2.4a,b,c)

The strains corresponding to Eqs. (2.1) - (2.4) are

$$\epsilon_{12}^{f(k,\ell)} = \frac{1}{2}\left(\partial_1\bar{u}_2^{(k,\ell)} + \psi_{21}^{f(k,\ell)}\right) \tag{2.5a}$$

$$\epsilon_{11}^{f(k,\ell)} = \bar{x}_2\,\partial_1\psi_{21}^{f(k,\ell)} \tag{2.5b}$$

$$\epsilon_{12}^{m(k,\ell)} = \frac{1}{2}\left(\partial_1\bar{u}_2^{(k,\ell)} + \psi_{21}^{m(k,\ell)}\right) \tag{2.6a}$$

$$\epsilon_{11}^{m(k,\ell)} = \bar{x}_2\,\partial_1\psi_{21}^{m(k,\ell)} \tag{2.6b}$$

or

$$\epsilon_{11}^{m(k,\ell)} = \pm\left(a^2-\bar{x}_3^2\right)^{\frac{1}{2}}\partial_1\psi_{21}^{f(k,\ell)} + \left[\bar{x}_2 \mp \left(a^2-\bar{x}_3^2\right)^{\frac{1}{2}}\right]\partial_1\psi_{21}^{m(k,\ell)}$$

(2.6c)

The strains $\epsilon_{13}^{m(k,\ell)}$ are neglected.

Next we will consider the conditions at the interface between cell (k,ℓ) and cell $(k+1,\ell)$, see Fig. 2.4. Since periodicity with respect to x_2 implies $\psi_{21}^{f(k,\ell)} = \psi_{21}^{f(k+1,\ell)}$, and $\psi_{21}^{m(k,\ell)} = \psi_{21}^{m(k+1,\ell)}$, we obtain by virtue of Eq. (2.5) of Part II:

$$\psi_{21}^{m(k,\ell)} = -\frac{\eta}{1-\eta}\,\psi_{21}^{f(k,\ell)}, \tag{2.7}$$

where η is the volume density of the fibers, i.e.,

$$\eta = \frac{A_f}{A_f + A_m} \tag{2.8}$$

The strains given by Eqs. (2.5a,b) and (2.6a,b,c) are now substituted in the expression for the strain energy, Eq. (1.4) of Part II, and the resulting expressions are integrated over the appropriate regions of cell (k,ℓ). The computation of the total strain energy averaged over the volume of cell (k,ℓ), as defined by Eq. (2.36) of Part II, and the subsequent transition to the continuum model, as described in Part II, then yield the following strain energy density:

$$W = \frac{1}{2} a_1 \left(\partial_1 \bar{u}_2\right)^2 + a_2 \left(\partial_1 \bar{u}_2\right) \psi^f_{21} + \frac{1}{2} a_3 \left(\psi^f_{21}\right)^2 + \frac{1}{2} a_4 \left(\partial_1 \psi^f_{21}\right)^2 \tag{2.9}$$

Here we have used the relation between ψ^m_{21} and ψ^f_{21} given by Eq. (2.7). The constants are:

$$a_1 = \eta\mu^f + (1-\eta)\mu^m \tag{2.10a}$$

$$a_2 = (\mu^f - \mu^m)\eta \tag{2.10b}$$

$$a_3 = \eta\mu^f + \frac{\eta^2}{1-\eta}\mu^m \tag{2.10c}$$

$$a_4 = 0.25\eta\,(\lambda_f + 2\mu_f)a^2 + (\lambda_m + 2\mu_m)C \tag{2.10d}$$

where

$$C = \left(\frac{4}{3}\frac{a^3}{d_3} - \frac{3}{4}\eta\, a^2\right)\left(1 + \frac{\eta}{1-\eta}\right)^2 - 2\left(\frac{1}{8}\eta d_2^2 - \frac{3}{8}\eta\, a^2\right)\left(1 + \frac{\eta}{1-\eta}\right)\frac{\eta}{1-\eta} + \left(\frac{1}{12}d_2^2 - \frac{1}{4}\eta\, a^2\right)\left(\frac{\eta}{1-\eta}\right)^2 \tag{2.10e}$$

The kinetic energy density is obtained as

$$T = \frac{1}{2}\bar{\rho}\left(\dot{\bar{u}}_2\right)^2 + \frac{1}{2} b\left(\dot{\psi}^f_{21}\right)^2 \tag{2.11}$$

where

$$\bar{\rho} = \eta\rho_f + (1-\eta)\rho_m \tag{2.12a}$$

$$b = 0.25\,\eta\,\rho_f a^2 + \rho_m C, \tag{2.12b}$$

and C is defined by Eq. (2.10e)

Application of the Euler-Poisson equation, given by Eq. (2.48) of Part II, where $F = T-W$ (the interface conditions have already been taken into account via Eq. (2.7)), yields

$$\bar{\rho}\,\ddot{\bar{u}}_2 - a_1\partial_1\partial_1\bar{u}_2 - a_2\partial_1\psi^f_{21} = 0 \tag{2.13}$$

$$b\ddot{\psi}_{21}^{f} - a_4\partial_1\partial_1\psi_{21}^{f} + a_2\partial_1\bar{u}_2 + a_3\psi_{21}^{f} = 0 \tag{2.14}$$

Let us consider harmonic waves of the forms

$$\left(\bar{u}_2,\psi_{21}^{f}\right) = \left(U_2,\, \Psi_{21}^{f}\right)e^{ik(x_1-ct)} \tag{2.15}$$

where $k = 2\pi/\Lambda$ is the wavenumber, Λ being the wavelength, and c is the phase velocity. Substitution of Eqs. (2.15) into (2.13) and (2.14) yields

$$\bar{\rho}c^2 = a_1 - \frac{a_2^2}{(a_4-bc^2)k^2 + a_3} \tag{2.16}$$

This is a quadratic equation for the phase velocity.

We will simplify the computation of c, by observing that for $k \to \infty$ we have $\bar{\rho}c^2 = a_1$. Since this upper limit is reached quickly, we substitute this result in the denominator, to obtain the explicit but approximate result:

$$\bar{\rho}c^2 = a_1 - \frac{a_2^2}{(a_4-ba_1/\bar{\rho})k^2 + a_3} \tag{2.17}$$

The group velocity, c_g, is related to the phase velocity by

$$c_g = c + k\,\frac{dc}{dk} \tag{2.18}$$

Table 3.3 Mechanical and geometric parameters of the boron-epoxy composite, Ref. 3.8.

Mechanical parameters	Boron	Epoxy (PR-279 resin)
Young's modulus in fiber direction, $E, 10^6$ psi	55.0	0.73
Mass density, ρ, $10^{-6}\ \frac{\text{lb}\ \text{sec}^2}{\text{in}^4}$	251	118
Poisson's ratio, ν, estimated	0.2	0.4
Geometric parameters		
fiber radius, in, a	0.002	
volume density, η,	0.54	
fiber radius/fiber spacing, a/d,	0.41	

Numerical results are presented for a boron-epoxy composite, for which experimental results were presented by Tauchert and Guzelsu, 3.8. The mechanical and geometrical parameters are summarized in Table 3.3. We use the same system of units as in Ref. 3.8. The values of the relevant ratios are

$$\gamma = \mu_f/\mu_m = 88.1 \quad ; \quad \theta = \rho_f/\rho_m = 2.13,$$

while

$$\left(c_T\right)_m = \left(\frac{\mu_m}{\rho_m}\right)^{\frac{1}{2}} = 0.469 \ \frac{\text{in}}{\mu\text{sec}}$$

The results computed from Eqs. (2.17) and (2.18) are plotted in Fig. 3.2. Experimental results from Ref. 3.8, Fig. 6, are also plotted in Fig. 3.2. It is noted that there are some deviations at small and relatively large frequencies, but the agreement is not altogether unsatisfactory.

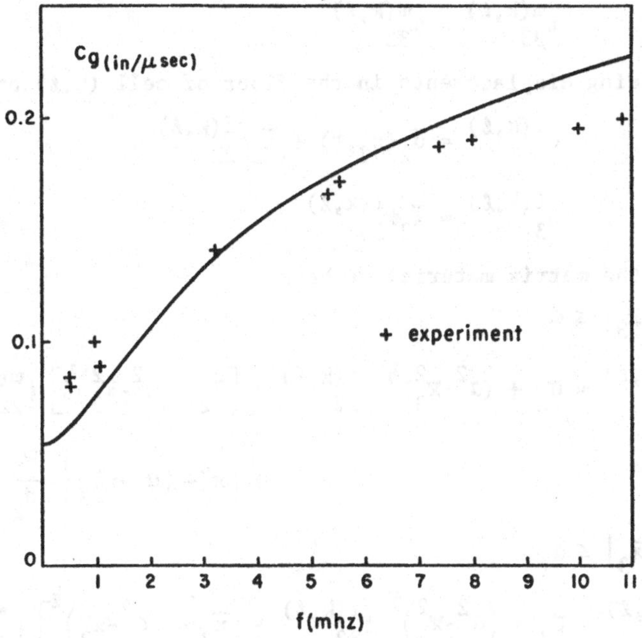

Fig. 3.2: Analytical and Experimental results for transverse waves propagating in the direction of the fibers; see Table 3.3 for mechanical and geometric parameters.

2.3 Longitudinal waves propagating normal to the fibers

Measurements of the phase velocity for various frequencies have been presented by, among others, Sutherland and Lingle, Ref. 3.9.

To construct a useful model for this more difficult case, it is necessary to include quadratic terms in the expansions inside cell (k,ℓ) and to provide an accurate representation of the interaction between neighboring cells.

Let us start off with expressions for the strains which are consistent with longitudinal motions in the x_2 direction. We consider

$$\epsilon_{22}^{f(k,\ell)} = \psi_{22}^{f(k,\ell)} \tag{2.19a}$$

$$\epsilon_{33}^{f(k,\ell)} = \psi_{33}^{f(k,\ell)} \tag{2.19b}$$

$$\epsilon_{22}^{m(k,\ell)} = \psi_{22}^{m(k,\ell)} + 2\,\frac{\bar{x}_2}{d_2}\,\varphi_{22} \tag{2.20a}$$

$$\epsilon_{33}^{m(k,\ell)} = \psi_{33}^{m(k,\ell)} \tag{2.20b}$$

Corresponding displacements in the fiber of cell (k,ℓ) are

$$u_2^{f(k,\ell)} = \bar{u}_2(x_2,t) + \bar{x}_2\psi_{22}^{f(k,\ell)}$$

$$u_3^{f(k,\ell)} = \bar{x}_3\psi_{33}^{f(k,\ell)}\ , \tag{2.21b}$$

while in the matrix material we have

$\bar{x}_2 \geq 0,\ |\bar{x}_3| \leq a$:

$$u_2^{m(k,\ell)} = \bar{u}_2 + (a^2-\bar{x}_3^2)^{\frac{1}{2}}\,\psi_{22}^{f(k,\ell)} + \left[\bar{x}_2 - \left(a^2-\bar{x}_3^2\right)^{\frac{1}{2}}\right]\psi_{22}^{m(k,\ell)} + \left[\bar{x}_2^2 - \left(a^2-\bar{x}_3^2\right)\right]\frac{\varphi_{22}}{d_2} \tag{2.22a}$$

$\bar{x}_2 \leq 0,\ |\bar{x}_3| \leq a$

$$u_2^{m(k,\ell)} = \bar{u}_2 - \left(a^2-\bar{x}_3^2\right)^{\frac{1}{2}}\psi_{22}^{f(k,\ell)} + \left[\bar{x}_2 + \left(a^2-\bar{x}_3^2\right)^{\frac{1}{2}}\right]\psi_{22}^{m(k,\ell)} + \left[\bar{x}_2^2 - \left(a^2-\bar{x}_3^2\right)\right]\frac{\varphi_{22}}{d_2} \tag{2.22b}$$

$a \leq |\bar{x}_3| \leq \frac{1}{2}\,d_3$

$$u_2^{m(k,\ell)} = \bar{u}_2 + \bar{x}_2\psi_{22}^{m(k,\ell)} + \bar{x}_2^2\,\frac{\varphi_{22}}{d_2} \tag{2.22c}$$

$\bar{x}_3 \geq 0,\ |\bar{x}_2| < a$:

$$u_3^{m(k,\ell)} = \left(a^2 - \bar{x}_2^2\right)^{\frac{1}{2}}\psi_{33}^{f(k,\ell)} + \left[\bar{x}_3 - \left(a^2 - \bar{x}_2^2\right)^{\frac{1}{2}}\right]\psi_{33}^{m(k,\ell)} \tag{2.22d}$$

$\bar{x}_3 \leq 0,\ |\bar{x}_2| < a$:

$$u_3^{m(k,\ell)} = -\left(a^2-\bar{x}_2^2\right)^{\frac{1}{2}}\psi_{33}^{f(k,\ell)} + \left[\bar{x}_3 + \left(a^2 - \bar{x}_2^2\right)^{\frac{1}{2}}\right]\psi_{33}^{m(k,\ell)} \tag{2.22e}$$

$a \leq |\bar{x}_2| \leq \frac{1}{2}\,d_2$:

$$u_3^{m(k,\ell)} = \bar{x}_3\psi_{33}^{m} \tag{2.22f}$$

These displacement distributions also give rise to shear strains

$\epsilon_{23}^{m(k,\ell)}$, which can be computed. It is evident that $\epsilon_{23}^{m(k,\ell)}$ assumes different values in different regions of the matrix material of cell (k,ℓ), and that the expressions for $\epsilon_{23}^{m(k,\ell)}$ depend on the local coordinatès. The averages $\bar{\epsilon}_{23}^{m(k,\ell)}$, over these regions are, however, easily computed. For example, for the region $a \le \bar{x}_3 \le \frac{1}{2} d_3$, $-a \le \bar{x}_2 \le 0$ we find

$$\bar{\epsilon}_{23}^{m(k,\ell)} = \frac{1}{2}\left(\psi_{33}^{f(k,\ell)} - \psi_{33}^{m(k,\ell)}\right) \tag{2.23}$$

while for $0 \le x_2 \le a$, $\left(a^2 - x_2^2\right)^{\frac{1}{2}} < x_3 < a$ we have

$$\bar{\epsilon}_{23}^{m(k,\ell)} = \frac{1}{4-\pi}\left(-\psi_{22}^{f(k,\ell)} + \psi_{22}^{m(k,\ell)} - \psi_{33}^{f(k,\ell)} + \psi_{33}^{m(k,\ell)}\right) \tag{2.24}$$

These expressions can be further simplified since $\psi_{22}^{f(k,\ell)}$ and $\psi_{33}^{f(k,\ell)}$ may be neglected as compared to $\psi_{22}^{m(k,\ell)}$ and $\psi_{33}^{m(k,\ell)}$ respectively.

Just as discussed in Part II, and exemplified in Section 2.2, the strains in the discrete cells lead us to the construction of a strain energy density. We find

$$W = \frac{1}{2} a_{22}^{f}\left(\psi_{22}^{f}\right)^2 + \frac{1}{2} a_{33}^{f}\left(\psi_{33}^{f}\right)^2 + a_{23}^{f}\psi_{22}^{f}\psi_{33}^{f} + \frac{1}{2} a_{22}^{m}\left(\psi_{22}^{m}\right)^2$$
$$+ \frac{1}{2} a_{33}^{m}\left(\psi_{33}^{m}\right)^2 + a_{23}^{m}\psi_{22}^{m}\psi_{33}^{m} + \frac{1}{2} b_{22}^{m}\varphi_{22}^{2} \tag{2.25}$$

where

$$a_{22}^{f} = a_{33}^{f} = \eta\left(\lambda_f + 2\mu_f\right)$$

$$a_{23}^{f} = \eta\lambda_f$$

$$a_{22}^{m} = \left(1-\eta\right)\left(\lambda_m + 2\mu_m\right) + \frac{2}{\pi}\,\frac{1}{4-\pi}\,\eta\mu_m + \left(\frac{d_2}{a} - 2\right)\frac{1}{\pi}\,\eta\mu_m$$

$$a_{33}^{m} = \left(1-\eta\right)\left(\lambda_m + 2\mu_m\right) + \frac{2}{\pi}\,\frac{1}{4-\pi}\,\eta\mu_m + \left(\frac{d_3}{a} - 2\right)\frac{1}{\pi}\,\eta\mu_m$$

$$a_{23}^{m} = \left(1-\eta\right)\lambda_m + \frac{4}{\pi}\,\frac{1}{4-\pi}\,\eta\mu_m$$

$$b_{22}^{m} = \left(\frac{1}{3} - \eta\,\frac{a^2}{d_2^2}\right)\left(\lambda_m + 2\mu_m\right) \tag{2.26}$$

In these expressions, η is defined by Eq. (2.8).

In the computation of the kinetic energy we only take into account the gross displacements, and we find

$$T = \frac{1}{2}\,\overline{\rho}\left(\dot{\overline{u}}_2\right)^2 \tag{2.27}$$

where $\overline{\rho}$ is defined by Eq. (2.12a)

The conditions at the interfaces of neighboring cells require careful consideration. Substituting the displacement expressions in Eq. (2.5) of Part II we find

$$\Delta_2\overline{u}_2 - p_3\,a\,\psi^f_{22} - \frac{1}{2}\,p_3 a\,\Delta_2\psi^f_{22} - \left(d_2 - p_3 a\right)\psi^m_{22} - \frac{1}{2}\left(d_2 - p_3 a\right)\Delta_2\psi^m_{22}$$

$$+ d_2\left(\frac{1}{4} - \frac{4}{3}\,\frac{a^3}{d_2^2 d_3}\right)\Delta_2\varphi_{22} = 0 \tag{2.28}$$

where

$$p_2 = \frac{\pi a}{d_2}\ , \quad p_3 = \frac{\pi a}{d_3} \tag{2.29a,b}$$

In a similar manner we find

$$S_3 = -\,p_2 a\psi^f_{33} - \left(d_3 - p_2 a\right)\psi^m_{33} = 0 \tag{2.30}$$

To place Eq. (2.28) within the context of the transition to the continuum model, we introduce Taylor expansions for the differences $\Delta_2 u$, etc. Defining the operator P[ ] as

$$P[\ \ ] = \sum_{n=1}^{\infty} \frac{1}{n!}\left(d_2\right)^n \frac{\partial^n}{\partial x_2^n} \tag{2.31}$$

we find that Eq. (2.28) can be replaced by

$$S_2 = P[\overline{u}_2] - p_3 a\psi^f_{22} - \frac{1}{2}\,p_3 a\,P[\psi^f_{22}] - \left(d_2 - p_3 a\right)\psi^m_{22}$$

$$-\,\frac{1}{2}\left(d_2 - p_3 a\right)P[\psi^m_{22}] + d_2\left(\frac{1}{4} - \frac{4}{3}\,\frac{a^3}{d_2^2 d_3}\right)P[\varphi_{22}] = 0 \tag{2.32}$$

The functional to be used in Hamilton's principle is now defined as

$$F = T - W - \lambda_2 S_2 - \lambda_3 S_3 \tag{2.33}$$

where λ_2 and λ_3 are Lagrangian multipliers. To obtain the appropriate

Euler-Poisson equation with F defined by Eq. (2.33), we employ the result given by Eq. (1.14):

$$\sum_{k=o}^{n_i} (-1)^k \frac{d^k}{dx^k} \left(\frac{\partial F}{\partial y_i^{(k)}}\right) = 0 \qquad (2.34)$$

The resulting system of equations is

$$- \bar{\rho}\, \ddot{\bar{u}}_2 - Q[\lambda_2] = 0$$

$$- a^f_{22}\psi^f_{22} - a^f_{23}\psi^f_{33} + p_3 a\, \lambda_2 + \frac{1}{2} p_3 a\, Q[\lambda_2] = 0$$

$$- a^m_{22}\psi^m_{22} - a^m_{23}\psi^m_{33} + (d_2 - p_3 a)\lambda_2 + \frac{1}{2}(d_2 - p_3 a)\, Q[\lambda_2] = 0$$

$$- a^f_{33}\psi^f_{33} - a^f_{23}\psi^f_{22} + p_2 a \lambda_3 = 0$$

$$- a^m_{33}\psi^m_{33} - a^m_{23}\psi^m_{22} + (d_3 - p_2 a)\lambda_2 = 0$$

$$-b^m_{22}\varphi_{22} - \frac{1}{4} d_2\, Q[\lambda_2] = 0 \qquad (2.35)$$

In these equations the operator Q[ ] is defined as

$$Q[\ ] = \sum_{n=1}^{\infty} \frac{1}{n!} (d_2)^n (-1)^n \frac{\partial^n}{\partial x_2^n} \qquad (2.36)$$

We will again consider expressions for the field variables representing harmonic waves, in this case propagating in the x_2-direction:

$$(\bar{u}_2, \lambda_2, \lambda_3, \varphi) = (U_2, \Lambda_2, \Lambda_3, \Phi) e^{ik(x_2 - ct)}$$

$$\left(\psi^f_{22}, \psi^f_{33}, \psi^m_{22}, \psi^m_{33}\right) = \left(\Psi^f_{22}, \Psi^f_{33}, \Psi^m_{22}, \Psi^m_{33}\right) e^{ik(x_2 - ct)}$$

Substitution of these expressions into Eqs. (2.35) and (2.30), (2.32) yields a system of eight homogeneous equations for the eight amplitudes: $U_2 \ldots \Psi^m_{33}$. The condition that the determinant must vanish yields an explicit expression for the frequency in terms of the wavenumber. We find

$$\Omega^2 = \frac{F(d_2 k)}{\eta\theta + 1 - \eta} \qquad (2.37)$$

where $\theta = \rho_f/\rho_m$, and the dimensionless frequency Ω is defined by

$$\Omega^2 = \frac{\omega^2 d_2^2}{\mu_m/\rho_m} \tag{2.38}$$

and η is defined by Eq. (2.8). The function F( ) which is a function of the dimensionless wavenumber d_2k is

$$F(d_2k) = \frac{1-\cos(d_2k)}{[1+\cos(d_2k)]\,M + [1-\cos(d_2k)]N} \tag{2.39}$$

where

$$M = \frac{0.25\ \eta[\eta D-(1-\eta)B] + 0.25(1-\eta)[1-\eta)A - \eta C]}{AD - BC}$$

$$N = \frac{0.0625(\lambda_m+2\mu_m)}{[0.333 - \eta p^2/d_2^2]\mu_m}$$

and

$$A = \left(a_{22}^f - \frac{(d_3-p_2a)^2}{p_2a}\frac{(a_{23}^f)^2}{E}\right)\frac{1}{\mu_m}$$

$$B = C = (d_3-p_2a)\,\frac{a_{23}^m a_{23}^f}{\mu_m E}$$

$$D = \left(a_{22}^m - \frac{p_2a(a_{23}^m)^2}{E}\right)\frac{1}{\mu_m}$$

$$E = \frac{(d_3-p_2a)^2}{p_2a}\,a_{33}^f + p_2a\,a_{33}^m$$

The function $F(d_2k)$ given by Eq. (2.39) implies a typical feature of wave propagation normal to the direction of the fibers, namely a maximum for Ω, with a corresponding stop band, and a value $d_2k = \pi$, i.e., Λ = wavelength = $2d_2$, at which the phase velocity vanishes.

Table 3.4 Mechanical and geometric parameters of the tungsten-aluminum composite, Ref. 3.9.

Mechanical parameters	Tungsten	Aluminum
Longitudinal modulus, $\lambda+2\mu$, dyne/cm^2	5.15×10^{12}	
Shear modulus, μ, dyne/cm^2		2.65×10^{11}
Poisson's ratio, ν,	0.28	0.34
Density, gm/cm^3	19.19	2.44 (22.1%)
		2.7 (2.2%)
Geometric parameters	2.2%	22.1%
volume density	0.022	0.221
fiber radius, mm	0.127	0.127
fiber radius/fiber spacing		
a/d_2	0.098	0.20
a/d_3	0.071	0.353

Experimental results for a tungsten-aluminum composite are presented in Ref. 3.9. The mechanical and geometrical parameters of this composite are summarized in Table 3.4. We use the same system of units as in Ref. 3.9. The phase velocity $c = \omega/k$ was computed from Eq. (2.37). In Fig. 3.3 the results are compared to the experimentally obtained values. Satisfactory agreement between theory and experiment was obtained.

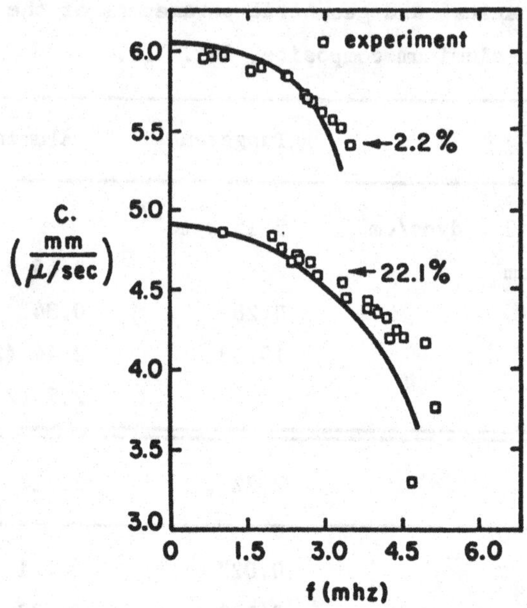

Fig. 3.3: Analytical and experimental results for longitudinal waves propagating normal to the fibers; see Table 3.4 for mechanical and geometric parameters.

2.4 <u>Longitudinal waves propagating along the fibers</u>

For longitudinal motions in the x_1-direction, the displacement distributions are symmetric with respect to the planes of structural symmetry of the fiber-reinforced composite. The case $d_2 = d_3 = d$, for which experimental information is available, has the additional simplifying feature that the dependence of the field variables on $\bar{x}_2$ is just the same as the dependence on $\bar{x}_3$.

Consistent with the foregoing observations we consider the following displacement distributions in the fiber of cell (k,ℓ).

(2.40a) $$u_1^{f(k,\ell)} = \bar{u}_1(x_1,t)$$

(2.40b) $$u_2^{f(k,\ell)} = \bar{x}_2\psi^{f(k,\ell)}(x_1,t)$$

(2.40c) $$u_3^{f(k,\ell)} = \bar{x}_3\psi^{f(k,\ell)}(x_1,t)$$

The corresponding strains are

$$\epsilon_{11}^{f(k,\ell)} = \partial_1 \bar{u}_1 \quad ; \quad \epsilon_{22}^{f(k,\ell)} = \epsilon_{33}^{f(k,\ell)} = \psi^{f(k,\ell)}$$

$$\epsilon_{12}^{f(k,\ell)} = \frac{1}{2} \bar{x}_2 \partial_1 \psi^{f(k,\ell)} \; ; \; \epsilon_{13}^{f(k,\ell)} = \frac{1}{2} \bar{x}_3 \, \partial_1 \psi^{f(k,\ell)} \tag{2.41}$$

For the displacements in the matrix material we choose

$$u_1^{m(k,\ell)} = \bar{u}_1(x_1,t) + d \sin\left(\frac{\pi}{2} \frac{r-a}{b-a}\right) \varphi(x_1,t) \tag{2.42}$$

where b is a radius such that

$$\pi b^2 = d^2 \tag{2.43}$$

Note that the displacements in the x_1-direction are continuous at $r = a$, and that the slope vanishes at $r = b$. The latter approximates the condition of displacement symmetry at the boundaries $\bar{x}_2 = \pm \frac{1}{2} d$ and $\bar{x}_3 = \pm \frac{1}{2} d$ of cell (k,ℓ). The displacements of the matrix material in the $\bar{x}_2$-and $\bar{x}_3$ directions are neglected. The strains in the matrix material follow as:

$$\epsilon_{11}^{m(k,\ell)} = \partial_1 \bar{u}_1 + d \sin\left(\frac{\pi}{2} \frac{r-a}{b-a}\right) \partial_1 \varphi \tag{2.44a}$$

$$\epsilon_{12}^{m(k,\ell)} = \frac{\pi}{4} \frac{d}{b-a} \cos\left(\frac{\pi}{2} \frac{r-a}{b-a}\right) \frac{\bar{x}_2}{r} \varphi \tag{2.45b}$$

$$\epsilon_{13}^{m(k,\ell)} = \frac{\pi}{4} \frac{d}{b-a} \cos\left(\frac{\pi}{2} \frac{r-a}{b-a}\right) \frac{\bar{x}_3}{r} \varphi \tag{2.45c}$$

The usual steps lead to the following strain energy density

$$W = \frac{1}{2} a_1 (\partial_1 \bar{u}_1)^2 + \frac{1}{2} a_2 (\psi^f)^2 + a_3 \partial_1 \bar{u}_1 \psi^f$$
$$+ \frac{1}{2} a_4 (\partial_1 \psi^f)^2 + a_5 \partial_1 \bar{u}_1 \partial_1 \varphi + \frac{1}{2} a_6 (\partial_1 \varphi)^2 + \frac{1}{2} a_7 \varphi^2 \tag{2.46}$$

where

$$a_1 = \eta(\lambda_f + 2\mu_f) + (1-\eta)(\lambda_m + 2\mu_m)$$

$$a_2 = 2\eta(\lambda_f + 2\mu_f) + 2\eta \lambda_f$$

$$a_3 = 2\eta \lambda_f$$

$$a_4 = \frac{1}{2} \eta a^2 \mu_f$$

$$a_5 = \left[\frac{8}{\pi^2}(1-\eta^{\frac{1}{2}})^2 + \frac{4}{\pi}\,\eta^{\frac{1}{2}}(1-\eta^{\frac{1}{2}})\right] d(\lambda_m + 2\mu_m)$$

$$a_6 = \left[\frac{8}{\pi^2}(1-\eta^{\frac{1}{2}})^2\left(\frac{\pi^2}{16} + \frac{1}{4}\right) + \eta^{\frac{1}{2}}(1-\eta^{\frac{1}{2}})\right] d^2\,(\lambda_m + 2\mu_m)$$

(2.47) $$a_7 = \left[\frac{\pi}{2}\left(\frac{\pi^2}{4} - 1\right) + \frac{\pi^3}{4}\,\eta^{\frac{1}{2}}(1-\eta^{\frac{1}{2}})^{-1}\right]\mu_m$$

For the computation of a kinetic energy density we consider the following particle velocities

(2.48a) $$\dot{\bar{u}}_1^{f(k,\ell)} = \dot{\bar{u}}_1$$

(2.48b) $$\dot{\bar{u}}_1^{m(k,\ell)} = \dot{\bar{u}}_1 + d\,\sin\left(\frac{\pi}{2}\,\frac{r-a}{b-a}\right)\varphi$$

We find

(2.49) $$T = \frac{1}{2}\,\bar{\rho}\,\dot{\bar{u}}^2 + b_1\dot{\bar{u}}_1\dot{\varphi} + \frac{1}{2}\,b_2\dot{\varphi}^2$$

where $\bar{\rho}$ is defined by Eq. (2.12a), and

(2.50a) $$b_1 = \left[\frac{8}{\pi^2}(1-\eta^{\frac{1}{2}})^2 + \frac{4}{\pi}\,\eta^{\frac{1}{2}}\,(1-\eta^{\frac{1}{2}})\right] d\,\rho_m$$

(2.50b) $$b_2 = \left[\frac{8}{\pi^2}\left(1-\eta^{\frac{1}{2}}\right)^2\left(\frac{\pi^2}{16} + \frac{1}{4}\right) + \eta^{\frac{1}{2}}\left(1-\eta^{\frac{1}{2}}\right)\right] d^2\rho_m$$

A straightforward application of Eq. (2.48) of Part II to F = T-W yields the following set of governing equations

(2.51) $$\bar{\rho}\,\ddot{\bar{u}}_1 + b_1\ddot{\varphi} - a_1\partial_1\partial_1\bar{u}_1 - a_3\partial_1\psi^f - a_5\partial_1\partial_1\varphi = 0$$

(2.52) $$a_2\psi^f + a_3\partial_1\bar{u}_1 - a_4\partial_1\partial_1\psi^f = 0$$

(2.53) $$b_1\ddot{\bar{u}}_1 + b_2\ddot{\varphi} - a_5\partial_1\partial_1\bar{u}_1 - a_6\partial_1\partial_1\varphi + a_7\varphi = 0$$

Substituting harmonic wave solutions of the forms

$$\left(u_1,\psi^f,\varphi\right) = \left(U,\Psi^f,\Phi\right)e^{ik(x_1-ct)}$$

in Eqs. (2.51) - (2.53), yields a relation between the phase velocity and the wavenumber as

(2.54) $$\bar{\rho}c^2 - a_1 + \frac{a_3^2}{a_2 + a_4k^2} - \frac{(b_1c^2 - a_5)^2k^2}{b_2c^2k^2 - a_6k^2 - a_7} = 0$$

This is a quadratic equation for c^2, which can easily be solved.

Table 3.5 Mechanical and geometric parameters of a fiber reinforced composite of silica fibers and polystyrene matrix material, Ref. 3.10.

Mechanical parameters	Silica	Polystyrene
Shear modulus, μ, dyne/cm^2	3.12×10^{11}	0.1323×10^{11}
Poisson's ratio, ν,	0.17	0.353
Density, gm/cm^3	2.2	1.056
Geometric parameters		
fiber radius, a, cm	0.051	
fiber spacing, d, cm	0.236	
volume density, $\eta = \pi a^2/d^2$	0.147	

In a recent article, Ref. 3.10, the finite element method was employed to investigate the dispersive characteristics of a fiber reinforced composite, for longitudinal motions propagating in the direction of the fibers. The computations were carried out for a composite whose mechanical and geometric parameters are summarized in Table 3.5. The results are shown in Fig. 3.4 by the solid line. The circles indicate experimental results presented in Ref. 3.10. For this composite the dimensionless phase velicity $c/(\mu_f/\rho_f)^{\frac{1}{2}}$ was computed versus $\frac{1}{2}$ kd from Eq. (2.54); the results are indicated in Fig. 3.4 by the dashed line.

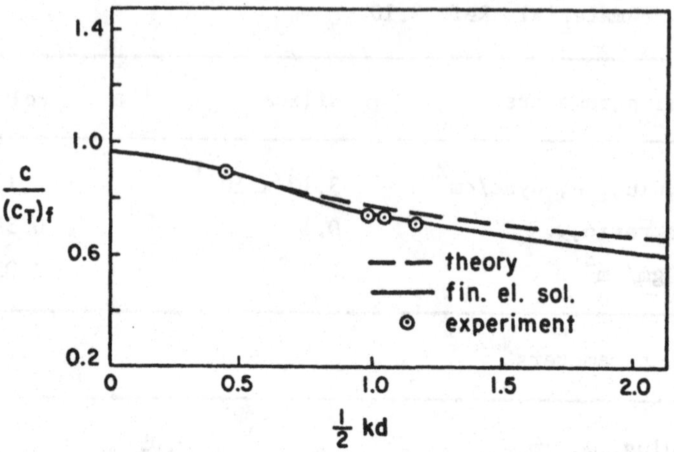

Fig. 3.4: Analytical, experimental, and numerical results for longitudinal waves propagating in the direction of the fibers; See Table 3.5 for mechanical and geometric parameters.

2.5 Concluding Remarks

In this chapter we have outlined a procedure to construct a generalized continuum theory for fiber-reinforced composites. For certain special wave motions, which are relevant to available data of ultrasonic tests on composite materials, equations governing the mechanical behavior were presented in detail, and analytical and experimental results were compared.

Within the framework of the theory presented here, the mechanical parameters of the constituents, and the geometric parameters describing the structuring of the composite, enter into coefficients in the set of governing partial differential equations. Thus the governing equations can be determined if relevant information on the constituents and the structuring of the composite is available. No unknown correction factors or other fudge devices, introduced for curve fitting purposes, enter in the theory presented here.

The potential applicability of generalized continuum theories to describe the mechanical behavior of composite materials was earlier discussed by Rivlin, Refs. 3.11 and 3.12, and Herrmann and Achenbach, Ref. 3.13. Alternative approaches to the one discussed here, have been presented by other authors. Among these we mention mixture theories, Refs. 3.14 and 3.15, and variational methods, Refs. 3.16 and 3.17.

It should be realized, of course, that the mechanical behavior of the constituents is not always known to the accuracy desired by theoreticians. In fact, due to manufacturing processes, the mechanical properties of the constituents may be somewhat different when part of a composite, as compared to the solitary state. In this light it seems to this writer that it is hardly necessary to require agreement on three digits accurate with "exact results". The agreement with experimental results presented here, which is in the five to ten percent range, is very satisfactory.

REFERENCES

3.1 S. M. Rytov, "Acoustical Properties of a Thinly Laminated Medium", Soviet Phys. Acoustics, Vol. 2, 1956, p. 68.

3.2 C.-T. Sun, J. D. Achenbach and G. Herrmann, "Time Harmonic Waves in a Stratified Medium, Propagating in the Direction of the Layering", J. Appl. Mech., Vol. 35, 1969, p. 408.

3.3 C. Sve, "Time-Harmonic Waves Traveling Obliquely in a Periodically Laminated Medium", J. Appl. Mech., Vol. 38, 1971, p. 477.

3.4 G. W. Postma, "Wave Propagation in a Stratified Medium", Geophysics, Vol. 20, 1955, p. 780.

3.5 C.-T. Sun, J. D. Achenbach and G. Herrmann, "Continuum Theory for a Laminated Medium", J. Appl. Mech., Vol. 35, 1968, p. 467.

3.6 L. Brillouin, Wave Propagation in Periodic Structures, Dover Publications, Inc., 1953.

3.7 C. W. Robinson and G. W. Leppelmeier, "Experimental Verification of Dispersion Relations for Layered Composites," J. Appl. Mech., Vol. 41, 1974, p. 89.

3.8 T. R. Tauchert and A. N. Guzelsu, "An Experimental Study of Dispersion of Stress Waves in a Fiber-Reinforced Composite", J. Appl. Mech., Vol. 39, 1972, p. 98.

3.9 H. J. Sutherland and R. Lingle, "Geometric Dispersion of Acoustic Waves by a Fibrous Composite", J. Composite Materials, Vol. 6, 1972, p. 490.

3.10 R. J. Talbot and J. S. Przemieniecki, "Finite Element Analysis of Frequency Spectra for Elastic Waveguides", Int. J. Solids and Structures, Vol. 11, 1975, p. 115.

3.11 R. S. Rivlin, "Generalized Mechanics of Continuous Media", Mechanics of Generalized Continuous Media, (E. Kröner, ed.), Springer Verlag, Berlin, 1968.

3.12 R. S. Rivlin, "The Formulation of Theories in Generalized Con tinuum Mechanics and their Physical Significance", Istituto Nazionale di Alta Mathematica, Symposia Mathematica, Vol. 1, 1968.

3.13 G. Herrmann and J. D. Achenbach, "Applications of Theories of Generalized Cosserat Continua to the Dynamics of Composite Materials", Mechanics of Generalized Continuous Media (E. Kröner, ed.). Springer Verlag, Berlin, 1968.

3.14 A. Bedford and M. Stern, "A Multi-Continuum Theory for Composite Elastic Materials", Acta Mechanica Vol. 14, 1972, p. 85.

3.15 G. A. Hegemier, "On a Theory of Interacting Continua for Wave Propagation in Composites," Dynamics of Composite Materials (E.H. Lee, ed.), The American Society of Mechanical Engineers, New York, 1972.

3.16 E. H. Lee, "A Survey of Variational Methods for Elastic Wave Propagation Analysis in Composites with Periodic Structures", Dynamics of Composite Materials (E. H. Lee, ed.), The American Society of Mechanical Engineers, New York, 1972.

3.17 S. Nemat-Nasser, F. C. L. Fu and S. Minagawa, "Harmonic Waves in One-, Two-and Three-dimensional Composites: Bounds for Eigenfrequencies", Int. J. Solids and Structures, Vol. 11, 1975, p. 617.

Printed in the United States
By Bookmasters